Essentials of Stem Cell Biology

Essentials of Stem Cell Biology

Edited by
Ian Cole

Larsen & Keller
www.larsen-keller.com

Essentials of Stem Cell Biology
Edited by Ian Cole
ISBN: 978-1-63549-268-2 (Hardback)

▤ Larsen & Keller

Published by Larsen and Keller Education,
5 Penn Plaza,
19th Floor,
New York, NY 10001, USA

Cataloging-in-Publication Data

Essentials of stem cell biology / edited by Ian Cole.
    p. cm.
Includes bibliographical references and index.
ISBN 978-1-63549-268-2
1. Stem cells. 2. Stem cells--Therapeutic use. 3. Cellular therapy. 4. Cells. 5. Cytology. I. Cole, Ian.
QH588.S8 E67 2017
616.027 74--dc23

The publisher's policy is to use permanent paper from mills that operate a sustainable forestry policy. Furthermore, the publisher ensures that the text paper and cover boards used have met acceptable environmental accreditation standards.

Printed and bound in the United States of America.

For more information regarding Larsen and Keller Education and its products, please visit the publisher's website www.larsen-keller.com

Table of Contents

Preface

This book is a valuable compilation of topics, ranging from the basic to the most complex theories and principles in the field of stem cell. It talks in detail about the various methods and applications of stem cell in our bodies. Stem cells refer to the undifferentiated biological cells present in multicellular organisms. These cells, through mitosis, can divide to produce many more stem cells. The text will give thorough insights into this subject area. The topics like embryonic stem cells and adult stem cells will be discussed in detail in the text. As this field is emerging at a rapid pace, the contents of the book will help the readers understand the modern concepts and applications of this subject.

Given below is the chapter wise description of the book:

Chapter 1- Stem cells are biological cells that split to form more stem cells. They can be located in organisms, which are multicellular. Stem cell lines are a group of stem cells that are nurtured in vitro and can be spread manually. The chapter on stem cell offers an insightful focus, keeping in mind the complex subject matter.

Chapter 2- The basic characteristic of a stem cell is its ability to regenerate tissues. The types of stem cells elucidated in this section are embryonic stem cells, adult stem cells, cancer stem cells, induced pluripotent stem cells, hematopoietic stem cells, mesenchymal stem cells, neural stem cells etc. Stem cells can best be understood in confluence with the major types listed in the following text.

Chapter 3- Stem-cell therapies are used to treat diseases. One of the most common forms of stem cell therapy is bone marrow transplant. In recent times, a lot of focus is put on the application of stem cell treatment for diseases such as diabetes and heart diseases. Many controversies related to stem cell therapy have also been discussed in this chapter. Some of these are human embryonic stem cells clinical trials, stem cell controversy and hematopoietic stem cell transplantation. This chapter is an overview of the subject matter incorporating all the major aspects of stem cell therapy.

Chapter 4- Bone marrow is the tissue found in the interior of the bones. There are two types of bone marrow, red marrow and yellow marrow. Red blood cells and most of the white blood cells are found in the red marrow. At birth all the bone marrow is red but with age it converts itself into the yellow type. This section will provide an integrated understanding of bone marrow.

Chapter 5- The significant aspects of stem cell are cell potency, cellular differentiation, epigenetics in stem-cell differentiation and stem cell laws. Cell potency is the cell's capability of changing from one cell type to another whereas embryonic stem cells are cells, which can regenerate and differentiate as per the requirements within the body. The aspects elucidated are of vital importance, and provide a better understanding of stem cells.

Indeed, my job was extremely crucial and challenging as I had to ensure that every chapter is informative and structured in a student-friendly manner. I am thankful for the support provided by my family and colleagues during the completion of this book.

Editor

Introduction to Stem Cell

Stem cells are biological cells that split to form more stem cells. They can be located in organisms, which are multicellular. Stem cell lines are a group of stem cells that are nurtured in vitro and can be spread manually. The chapter on stem cell offers an insightful focus, keeping in mind the complex subject matter.

Stem Cell

Stem cells are undifferentiated biological cells that can differentiate into specialized cells and can divide (through mitosis) to produce more stem cells. They are found in multicellular organisms. In mammals, there are two broad types of stem cells: embryonic stem cells, which are isolated from the inner cell mass of blastocysts, and adult stem cells, which are found in various tissues. In adult organisms, stem cells and progenitor cells act as a repair system for the body, replenishing adult tissues. In a developing embryo, stem cells can differentiate into all the specialized cells—ectoderm, endoderm and mesoderm but also maintain the normal turnover of regenerative organs, such as blood, skin, or intestinal tissues.

There are three known accessible sources of autologous adult stem cells in humans:

1. Bone marrow, which requires extraction by *harvesting*, that is, drilling into bone (typically the femur or iliac crest).

2. Adipose tissue (lipid cells), which requires extraction by liposuction.

3. Blood, which requires extraction through apheresis, wherein blood is drawn from the donor (similar to a blood donation), and passed through a machine that extracts the stem cells and returns other portions of the blood to the donor.

Stem cells can also be taken from umbilical cord blood just after birth. Of all stem cell types, autologous harvesting involves the least risk. By definition, autologous cells are obtained from one's own body, just as one may bank his or her own blood for elective surgical procedures.

Adult stem cells are frequently used in various medical therapies (e.g., bone marrow transplantation). Stem cells can now be artificially grown and transformed (differentiated) into specialized cell types with characteristics consistent with cells of various tissues such as muscles or nerves. Embryonic cell lines and autologous embryonic stem cells generated through somatic cell nuclear transfer or dedifferentiation have also been proposed as promising candidates for future therapies. Research into stem cells grew out of findings by Ernest A. McCulloch and James E. Till at the University of Toronto in the 1960s.

Properties

The classical definition of a stem cell requires that it possess two properties:

- *Self-renewal*: the ability to go through numerous cycles of cell division while maintaining the undifferentiated state.

- *Potency*: the capacity to differentiate into specialized cell types. In the strictest sense, this requires stem cells to be either totipotent or pluripotent—to be able to give rise to any mature cell type, although multipotent or unipotent progenitor cells are sometimes referred to as stem cells. Apart from this it is said that stem cell function is regulated in a feed back mechanism.

Self-renewal

Two mechanisms exist to ensure that a stem cell population is maintained:

1. Obligatory asymmetric replication: a stem cell divides into one mother cell that is identical to the original stem cell, and another daughter cell that is differentiated.

2. Stochastic differentiation: when one stem cell develops into two differentiated daughter cells, another stem cell undergoes mitosis and produces two stem cells identical to the original.

Potency Definition

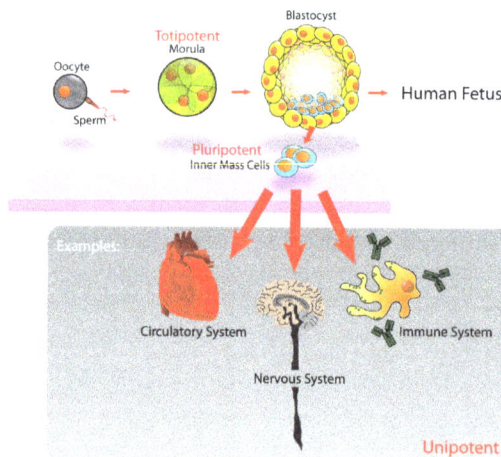

Pluripotent, embryonic stem cells originate as inner cell mass (ICM) cells within a blastocyst. These stem cells can become any tissue in the body, excluding a placenta. Only cells from an earlier stage of the embryo, known as the morula, are totipotent, able to become all tissues in the body and the extraembryonic placenta.

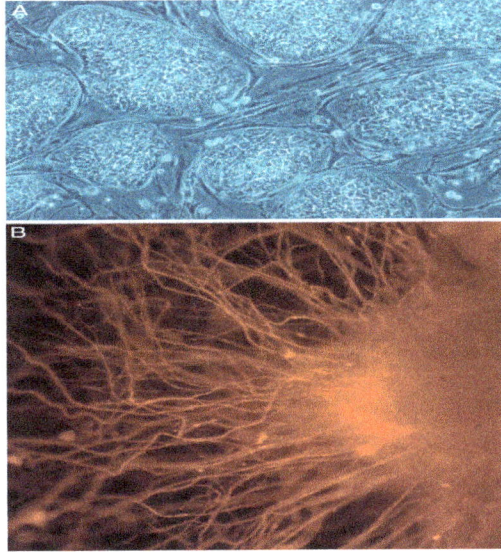

Human embryonic stem cells

A: Stem cell colonies that are not yet differentiated.

B: Nerve cells, an example of a cell type after differentiation.

Potency specifies the differentiation potential (the potential to differentiate into different cell types) of the stem cell.

- Totipotent (a.k.a. omnipotent) stem cells can differentiate into embryonic and extraembryonic cell types. Such cells can construct a complete, viable organism. These cells are produced from the fusion of an egg and sperm cell. Cells produced by the first few divisions of the fertilized egg are also totipotent.

- Pluripotent stem cells are the descendants of totipotent cells and can differentiate into nearly all cells, i.e. cells derived from any of the three germ layers.

- Multipotent stem cells can differentiate into a number of cell types, but only those of a closely related family of cells.

- Oligopotent stem cells can differentiate into only a few cell types, such as lymphoid or myeloid stem cells.

- Unipotent cells can produce only one cell type, their own, but have the property of self-renewal, which distinguishes them from non-stem cells (e.g. progenitor cells, which cannot self-renew).

Identification

In practice, stem cells are identified by whether they can regenerate tissue. For example, the defining test for bone marrow or hematopoietic stem cells (HSCs) is the ability to transplant the cells and save an individual without HSCs. This demonstrates that the cells can produce new blood cells over a long term. It should also be possible to isolate stem cells from the transplanted individual, which can themselves be transplanted into another individual without HSCs, demonstrating that the stem cell was able to self-renew.

Properties of stem cells can be illustrated *in vitro*, using methods such as clonogenic assays, in which single cells are assessed for their ability to differentiate and self-renew. Stem cells can also be isolated by their possession of a distinctive set of cell surface markers. However, *in vitro* culture conditions can alter the behavior of cells, making it unclear whether the cells will behave in a similar manner *in vivo*. There is considerable debate as to whether some proposed adult cell populations are truly stem cells.

Embryonic

Embryonic stem (ES) cells are the cells of the inner cell mass of a blastocyst, an early-stage embryo. Human embryos reach the blastocyst stage 4–5 days post fertilization, at which time they consist of 50–150 cells. ES cells are pluripotent and give rise during development to all derivatives of the three primary germ layers: ectoderm, endoderm and mesoderm. In other words, they can develop into each of the more than 200 cell types of the adult body when given sufficient and necessary stimulation for a specific cell type. They do not contribute to the extra-embryonic membranes or the placenta.

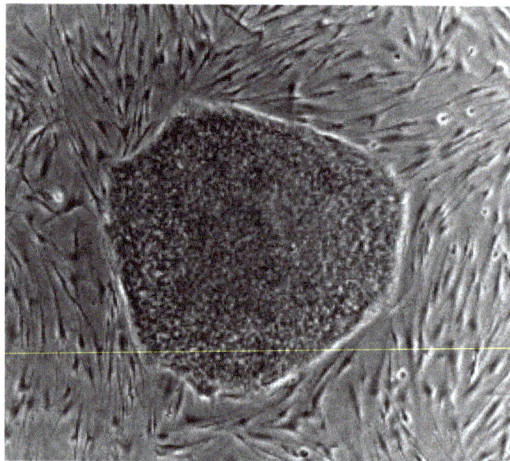

Human embryonic stem cell colony on mouse embryonic fibroblast feeder layer

During embryonic development these inner cell mass cells continuously divide and become more specialized. For example, a portion of the ectoderm in the dorsal part of the embryo specializes as 'neurectoderm', which will become the future central nervous system. Later in development, neurulation causes the neurectoderm to form the neural tube. At the neural tube stage, the anterior portion undergoes encephalization to generate or 'pattern' the basic form of the brain. At this stage of development, the principal cell type of the CNS is considered a neural stem cell. These neural stem cells are pluripotent, as they can generate a large diversity of many different neuron types, each with unique gene expression, morphological, and functional characteristics. One prominent example of a neural stem cell is the radial glial cell, so named because it has a distinctive bipolar morphology with highly elongated processes spanning the thickness of the neural tube wall, and because historically it shared some glial characteristics, most notably the expression of glial fibrillary acidic protein (GFAP). The radial glial cell is the primary neural stem cell of the developing vertebrate CNS, and its cell body resides in the ventricular zone, adjacent to the developing ventricular system. Neural stem cells are committed to the neuronal lineages (neurons, astrocytes, and oligodendrocytes), and thus their potency is restricted.

Nearly all research to date has made use of mouse embryonic stem cells (mES) or human embryonic stem cells (hES) derived from the early inner cell mass. Both have the essential stem cell characteristics, yet they require very different environments in order to maintain an undifferentiated state. Mouse ES cells are grown on a layer of gelatin as an extracellular matrix (for support) and require the presence of leukemia inhibitory factor (LIF). Human ES cells are grown on a feeder layer of mouse embryonic fibroblasts (MEFs) and require the presence of basic fibroblast growth factor (bFGF or FGF-2). Without optimal culture conditions or genetic manipulation, embryonic stem cells will rapidly differentiate.

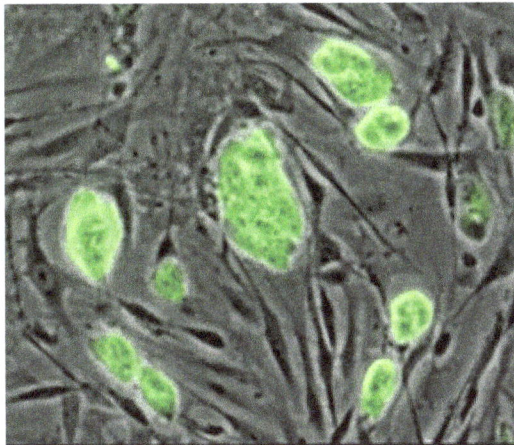

Mouse embryonic stem cells with fluorescent marker

A human embryonic stem cell is also defined by the expression of several transcription factors and cell surface proteins. The transcription factors Oct-4, Nanog, and Sox2 form the core regulatory network that ensures the suppression of genes that lead to differentiation and the maintenance of pluripotency. The cell surface antigens most commonly used to identify hES cells are the glycolipids stage specific embryonic antigen 3 and 4 and the keratan sulfate antigens Tra-1-60 and Tra-1-81. By using human embryonic stem cells to produce specialized cells like nerve cells or heart cells in the lab, scientists can gain access to adult human cells without taking tissue from patients. They can then study these specialized adult cells in detail to try and catch complications of diseases, or to study cells reactions to potentially new drugs. The molecular definition of a stem cell includes many more proteins and continues to be a topic of research.

There are currently no approved treatments using embryonic stem cells. The first human trial was approved by the US Food and Drug Administration in January 2009. However, the human trial was not initiated until October 13, 2010 in Atlanta for spinal cord injury research. On November 14, 2011 the company conducting the trial (Geron Corporation) announced that it will discontinue further development of its stem cell programs. ES cells, being pluripotent cells, require specific signals for correct differentiation—if injected directly into another body, ES cells will differentiate into many different types of cells, causing a teratoma. Differentiating ES cells into usable cells while avoiding transplant rejection are just a few of the hurdles that embryonic stem cell researchers still face. Due to ethical considerations, many nations currently have moratoria or limitations on either human ES cell research or the production of new human ES cell lines. Because of their combined abilities of unlimited expansion and pluripotency, embryonic stem cells remain a theoretically potential source for regenerative medicine and tissue replacement after injury or disease.

Fetal

The primitive stem cells located in the organs of fetuses are referred to as fetal stem cells. There are two types of fetal stem cells:

1. Fetal proper stem cells come from the tissue of the fetus proper, and are generally obtained after an abortion. These stem cells are not immortal but have a high level of division and are multipotent.

2. Extraembryonic fetal stem cells come from extraembryonic membranes, and are generally not distinguished from adult stem cells. These stem cells are acquired after birth, they are not immortal but have a high level of cell division, and are pluripotent.

Adult

Adult stem cells, also called somatic (from Greek "of the body") stem cells, are stem cells which maintain and repair the tissue in which they are found. They can be found in children, as well as adults.

Pluripotent adult stem cells are rare and generally small in number, but they can be found in umbilical cord blood and other tissues. Bone marrow is a rich source of adult stem cells, which have been used in treating several conditions including spinal cord injury, liver cirrhosis, chronic limb ischemia and endstage heart failure. The quantity of bone marrow stem cells declines with age and is greater in males than females during reproductive years. Much adult stem cell research to date has aimed to characterize their potency and self-renewal capabilities. DNA damage accumulates with age in both stem cells and the cells that comprise the stem cell environment. This accumulation is considered to be responsible, at least in part, for increasing stem cell dysfunction with aging.

Most adult stem cells are lineage-restricted (multipotent) and are generally referred to by their tissue origin (mesenchymal stem cell, adipose-derived stem cell, endothelial stem cell, dental pulp stem cell, etc.).

Adult stem cell treatments have been successfully used for many years to treat leukemia and related bone/blood cancers through bone marrow transplants. Adult stem cells are also used in veterinary medicine to treat tendon and ligament injuries in horses.

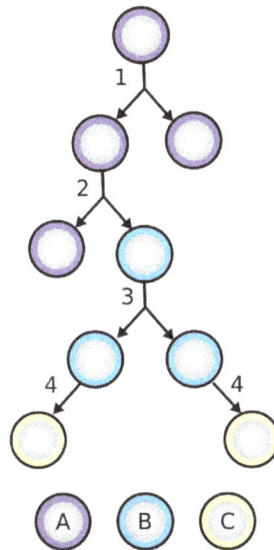

Stem cell division and differentiation. A: stem cell; B: progenitor cell; C: differentiated cell; 1: symmetric stem cell division; 2: asymmetric stem cell division; 3: progenitor division; 4: terminal differentiation

The use of adult stem cells in research and therapy is not as controversial as the use of embryonic stem cells, because the production of adult stem cells does not require the destruction of an embryo. Additionally, in instances where adult stem cells are obtained from the intended recipient (an autograft), the risk of rejection is essentially non-existent. Consequently, more US government funding is being provided for adult stem cell research.

Amniotic

Multipotent stem cells are also found in amniotic fluid. These stem cells are very active, expand extensively without feeders and are not tumorigenic. Amniotic stem cells are multipotent and can differentiate in cells of adipogenic, osteogenic, myogenic, endothelial, hepatic and also neuronal lines. Amniotic stem cells are a topic of active research.

Use of stem cells from amniotic fluid overcomes the ethical objections to using human embryos as a source of cells. Roman Catholic teaching forbids the use of embryonic stem cells in experimentation; accordingly, the Vatican newspaper "Osservatore Romano" called amniotic stem cells "the future of medicine".

It is possible to collect amniotic stem cells for donors or for autologuous use: the first US amniotic stem cells bank was opened in 2009 in Medford, MA, by Biocell Center Corporation and collaborates with various hospitals and universities all over the world.

Induced Pluripotent

These are not adult stem cells, but rather adult cells (e.g. epithelial cells) reprogrammed to give rise to pluripotent capabilities. Using genetic reprogramming with protein transcription factors, pluripotent stem cells equivalent to embryonic stem cells have been derived from human adult skin tissue. Shinya Yamanaka and his colleagues at Kyoto University used the transcription factors Oct3/4, Sox2, c-Myc, and Klf4 in their experiments on human facial skin cells. Junying Yu, James Thomson, and their colleagues at the University of Wisconsin–Madison used a different set of factors, Oct4, Sox2, Nanog and Lin28, and carried out their experiments using cells from human foreskin.

As a result of the success of these experiments, Ian Wilmut, who helped create the first cloned animal Dolly the Sheep, has announced that he will abandon somatic cell nuclear transfer as an avenue of research.

Frozen blood samples can be used as a source of induced pluripotent stem cells, opening a new avenue for obtaining the valued cells.

Lineage

To ensure self-renewal, stem cells undergo two types of cell division. Symmetric division gives rise to two identical daughter cells both endowed with stem cell properties. Asymmetric division, on the other hand, produces only one stem cell and a progenitor cell with limited self-renewal potential. Progenitors can go through several rounds of cell division before terminally differentiating into a mature cell. It is possible that the molecular distinction between symmetric and asymmetric divisions lies in differential segregation of cell membrane proteins (such as receptors) between the daughter cells.

An alternative theory is that stem cells remain undifferentiated due to environmental cues in their particular niche. Stem cells differentiate when they leave that niche or no longer receive those signals. Studies in *Drosophila* germarium have identified the signals decapentaplegic and adherens junctions that prevent germarium stem cells from differentiating.

Treatments

Potential uses of
Stem cells

Stroke | Baldness
Traumatic brain injury | Blindness
Learning defects | Deafness
Alzheimer's disease | Amyotrophic lateral-sclerosis
Parkinson's disease
Missing teeth | Myocardial infarction
Wound healing | Muscular dystrophy
Bone marrow transplantation (currently established) | Diabetes
Spinal cord injury | Multiple sites: Cancers
Osteoarthritis
Rheumatoid arthritis | Crohn's disease

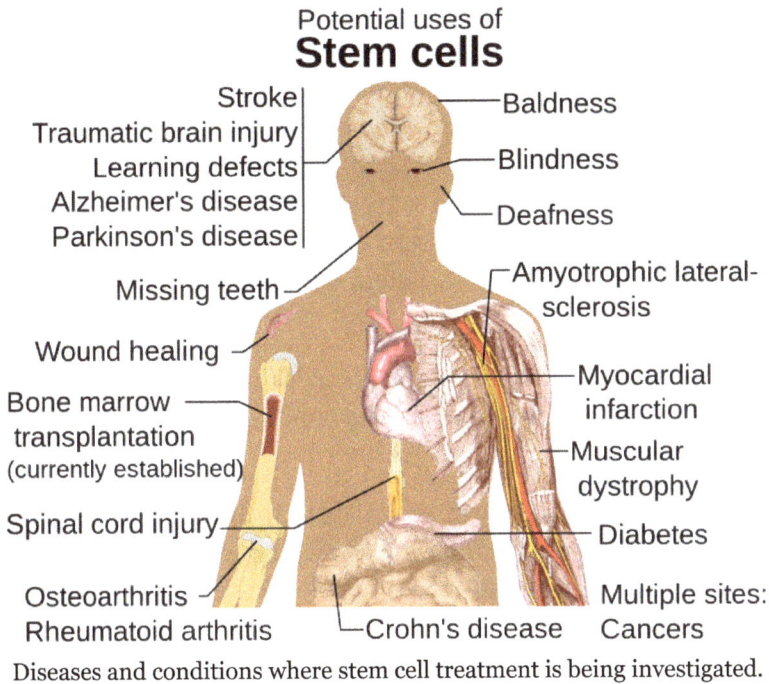

Diseases and conditions where stem cell treatment is being investigated.

Diseases and conditions where stem cell treatment is being investigated include:

- Diabetes

- Rheumatoid arthritis

- Parkinson's disease

- Alzheimer's disease

- Osteoarthritis

- Stroke and traumatic brain injury repair

- Learning disability due to congenital disorder

- Spinal cord injury repair

- Heart infarction

- Anti-cancer treatments

- Baldness reversal

- Replace missing teeth

- Repair hearing

- Restore vision and repair damage to the cornea

- Amyotrophic lateral sclerosis

- Crohn's disease

- Wound healing

Stem cell therapy is the use of stem cells to treat or prevent a disease or condition. Bone marrow transplant is a crude form of stem cell therapy that has been used clinically for many years without controversy. No stem cell therapies other than bone marrow transplant are widely used.

Research is underway to develop various sources for stem cells, and to apply stem cell treatments for neurodegenerative diseases and conditions, diabetes, heart disease, and other conditions.

In more recent years, with the ability of scientists to isolate and culture embryonic stem cells, and with scientists' growing ability to create stem cells using somatic cell nuclear transfer and techniques to create induced pluripotent stem cells, controversy has crept in, both related to abortion politics and to human cloning.

Hepatotoxicity and drug-induced liver injury account for a substantial number of failures of new drugs in development and market withdrawal, highlighting the need for screening assays such as stem cell-derived hepatocyte-like cells, that are capable of detecting toxicity early in the drug development process.

Disadvantages

Stem cell treatments may require immunosuppression because of a requirement for radiation before the transplant to remove the patient's previous cells, or because the patient's immune system may target the stem cells. One approach to avoid the second possibility is to use stem cells from the same patient who is being treated.

Pluripotency in certain stem cells could also make it difficult to obtain a specific cell type. It is also difficult to obtain the exact cell type needed, because not all cells in a population differentiate uniformly. Undifferentiated cells can create tissues other than desired types.

Some stem cells form tumors after transplantation; pluripotency is linked to tumor formation especially in embryonic stem cells, fetal proper stem cells, induced pluripotent stem cells. Fetal proper stem cells form tumors despite multipotency.

Research Patents

Some of the fundamental patents covering human embryonic stem cells are owned by the Wisconsin Alumni Research Foundation (WARF) - they are patents 5,843,780,

6,200,806, and 7,029,913 invented by James A. Thomson. WARF does not enforce these patents against academic scientists, but does enforce them against companies.

In 2006, a request for the US Patent and Trademark Office (USPTO) to re-examine the three patents was filed by the Public Patent Foundation on behalf of its client, the non-profit patent-watchdog group Consumer Watchdog (formerly the Foundation for Taxpayer and Consumer Rights). In the re-examination process, which involves several rounds of discussion between the USTPO and the parties, the USPTO initially agreed with Consumer Watchdog and rejected all the claims in all three patents, however in response, WARF amended the claims of all three patents to make them more narrow, and in 2008 the USPTO found the amended claims in all three patents to be patentable. The decision on one of the patents (7,029,913) was appealable, while the decisions on the other two were not. Consumer Watchdog appealed the granting of the '913 patent to the USTPO's Board of Patent Appeals and Interferences (BPAI) which granted the appeal, and in 2010 the BPAI decided that the amended claims of the '913 patent were not patentable. However, WARF was able to re-open prosecution of the case and did so, amending the claims of the '913 patent again to make them more narrow, and in January 2013 the amended claims were allowed.

In July 2013, Consumer Watchdog announced that it would appeal the decision to allow the claims of the '913 patent to the US Court of Appeals for the Federal Circuit (CAFC), the federal appeals court that hears patent cases. At a hearing in December 2013, the CAFC raised the question of whether Consumer Watchdog had legal standing to appeal; the case could not proceed until that issue was resolved.

Stem-cell Line

A stem-cell line is a group of stem cells that is cultured in vitro and can be propagated indefinitely. Stem-cell lines are derived from either animal or human tissues and come from one of three sources: embryonic stem cells, adult stem cells, or induced stem cells. They are commonly used in research and regenerative medicine.

Properties

By definition, stem cells possess two properties: (1) they can self-renew, which means that they can divide indefinitely while remaining in an undifferentiated state; and (2) they are pluripotent or multipotent, which means that they can differentiate to form specialized cell types. Due to the self-renewal capacity of stem cells, a stem cell line can be cultured *in vitro* indefinitely.

A stem-cell line is distinctly different from an immortalized cell line, such as the HeLa line. While stem cells can propagate indefinitely in culture due to their inherent properties, immortalized cells would not normally divide indefinitely but have gained this ability due to mutation. Immortalized cell lines can be generated from cells isolated from tumors, or mutations can be introduced to make the cells immortal.

A stem cell line is also distinct from primary cells. Primary cells are cells that have been isolated and then used immediately. Primary cells cannot divide indefinitely and thus cannot be cultured for long periods of time in vitro.Types and methods of derivation

Embryonic Stem Cell Line

An embryonic stem cell line is created from cells derived from the inner cell mass of a blastocyst, an early stage, pre-implantation embryo. In humans, the blastocyst stage occurs 4–5 days post fertilization. To create an embryonic stem cell line, the inner cell-mass is removed from the blastocyst, separated from the trophoectoderm, and cultured on a layer of supportive cells in vitro. In the derivation of human embryonic stem cell lines, embryos leftover from in vitro fertilization (IVF) procedures are used. The fact that the blastocyst is destroyed during the process has raised controversy and ethical concerns.

Embryonic stem cells are pluripotent, meaning they can differentiate to form all cell types in the body. In vitro, embryonic stem cells can be cultured under defined conditions to keep them in their pluripotent state, or they can be stimulated with biochemical and physical cues to differentiate them to different cell types.

Adult Stem Cell Line

Adult stem cells are found in juvenile or adult tissues. Adult stem cells are multipotent: they can generate a limited number of differentiated cell types (unlike pluripotent embryonic stem cells). Types of adult stem cells include hematopoietic stem cells and mesenchymal stem cells. Hematopoetic stem cells are found in the bone marrow and generate all cells of the immune system all blood cell types. Mesenchymal stem cells are found in umbilical cord blood, amniotic fluid, and adipose tissue and can generate a number of cell types, including osteoblasts, chondrocytes, and adipocytes. In medicine, adult stem cells are mostly commonly used in bone marrow transplants to treat many bone and blood cancers as well as some autoimmune diseases.

Of the types of adult stem cells have successfully been isolated and identified, only mesenchymal stem cells can successfully be grown in culture for long periods of time. Other adult stem cell types, such as hematopoietic stem cells, are difficult to grow and propagate in vitro. Identifying methods for maintaining hematopoietic stem cells in vitro is an active area of research. Thus, while mesenchymal stem cell lines exist, other types of adult stem cells that are grown in vitro can better be classified as primary cells.

Induced Pluripotent Stem-cell (IPSC) Line

Induced pluripotent stem cell (iPSC) lines are pluripotent stem cells that have been generated from adult/somatic cells. The method of generating iPSCs was developed by Shinya Yamanaka's lab in 2006; his group demonstrated that the introduction of four specific genes could induce somatic cells to revert to a pluripotent stem cell state.

Compared to embryonic stem-cell lines, iPSC lines are also pluripotent in nature but can be derived without the use of human embryos—a process that has raised ethical concerns. Furthermore, patient-specific iPSC cell lines can be generated—that is, cell lines that are genetically matched to an individual. Patient-specific iPSC lines have been generated for the purposes of studying diseases and for developing patient-specific medical therapies.

Methods of Culture

Stem-cell lines are grown and maintained at specific temperature and atmospheric conditions (37 degrees Celsius and 5% CO_2) in incubators. Culture conditions such as the cell growth medium and surface on which cells are grown vary widely depending on the specific stem cell line. Different biochemical factors can be added to the medium to control the cell phenotype—for example to keep stem cells in a pluripotent state or to differentiate them to a specific cell type.

Uses

Stem-cell lines are used in research and regenerative medicine. They can be used to study stem-cell biology and early human development. In the field of regenerative medicine, it has been proposed that stem cells be used in cell-based therapies to replace injured or diseased cells and tissues. Examples of conditions that researchers are working to develop stem-cell-based treatments for include neurodegenerative diseases, diabetes, and spinal cord injuries.

Ethical Issues

There is controversy associated with the derivation and use of human embryonic stem cell lines. This controversy stems from the fact that derivation of human embryonic stem cells requires the destruction of a blastocyst-stage, pre-implantation human embryo. There is a wide range of viewpoints regarding the moral consideration that blastocyst-stage human embryos should be given.

Access to Human Embryonic Stem-cell Lines

United States

In the United States, Executive Order 13505 established that federal money can be used for research in which approved human embryonic stem-cell (hESC) lines are used, but it cannot be used to derive new lines. The National Institutes of Health (NIH) Guidelines on Human Stem Cell Research, effective July 7, 2009, implemented the Executive Order 13505 by establishing criteria which hESC lines must meet to be approved for funding. The NIH Human Embryonic Stem Cell Registry can be accessed online and has updated information on cell lines eligible for NIH funding. There are 279 approved lines as of April 2014.

Studies have found that approved hESC lines are not uniformly used in the US data from cell banks and surveys of researchers indicate that only a handful of the available hESC lines are routinely used in research. Access and utility are cited as the two primary factors influencing what hESC lines scientists choose to work with.

A 2011 survey of stem cell scientists in the US who use hESC lines in their research found that 54% of respondents used two or fewer lines and 75% used three or fewer lines.

Another study tracked cell-line requests fulfilled from the largest US repositories, the National Stem Cell Bank (NSCB) and the Harvard Stem Cell Institute (HSCI; Cambridge, MA, USA), for the periods March 1999 – December 2008 (for NSCB) and April 2004 – December 2008 (for HSCI). For NSCB, out of twenty-one approved cell lines, 77% of requests were for two of the lines (H1 and H9). For HSCI, out of the 17 lines requested more than once, 24.7% of requests were for the two most commonly requested lines.

Stem-cell Niche

Stem-cell niche refers to a microenvironment, within the specific anatomic location where stem cells are found, which interacts with stem cells to regulate cell fate. The word 'niche' can be in reference to the *in vivo* or *in vitro* stem-cell microenvironment. During embryonic development, various niche factors act on embryonic stem cells to alter gene expression, and induce their proliferation or differentiation for the development of the fetus. Within the human body, stem-cell niches maintain adult stem cells in a quiescent state, but after tissue injury, the surrounding micro-environment actively signals to stem cells to promote either self-renewal or differentiation to form new tissues. Several factors are important to regulate stem-cell characteristics within the niche: cell–cell interactions between stem cells, as well as interactions between stem cells and neighbouring differentiated cells, interactions between stem cells and adhesion molecules, extracellular matrix components, the oxygen tension, growth factors, cytokines, and the physicochemical nature of the environment including the pH, ionic strength (e.g. Ca^{2+} concentration) and metabolites, like ATP, are also important. The stem cells and niche may induce each other during development and reciprocally signal to maintain each other during adulthood.

Scientists are studying the various components of the niche and trying to replicate the *in vivo* niche conditions *in vitro*. This is because for regenerative therapies, cell proliferation and differentiation must be controlled in flasks or plates, so that sufficient quantity of the proper cell type are produced prior to being introduced back into the patient for therapy.

Human embryonic stem cells are often grown in fibroblastic growth factor-2 containing, fetal bovine serum supplemented media. They are grown on a feeder layer of cells, which is believed to be supportive in maintaining the pluripotent characteristics of embryonic stem cells. However, even these conditions may not truly mimic *in vivo* niche conditions.

Adult stem cells remain in an undifferentiated state throughout adult life. However, when they are cultured *in vitro*, they often undergo an 'aging' process in which their morphology is changed and their proliferative capacity is decreased. It is believed that correct culturing conditions of adult stem cells needs to be improved so that adult stem cells can maintain their stemness over time.A *Nature* Insight review defines niche as follows:

"Stem-cell populations are established in 'niches' — specific anatomic locations that regulate how they participate in tissue generation, maintenance and repair. The niche saves stem cells from depletion, while protecting the host from over-exuberant stem-cell proliferation. It constitutes a basic unit of tissue physiology, integrating signals that mediate the balanced response of stem cells to the needs of organisms. Yet the niche may also induce pathologies by imposing aberrant function on stem cells or other targets. The interplay between stem cells and their niche creates the dynamic system necessary for sustaining tissues, and for the ultimate design of stem-cell therapeutics ... The simple location of stem cells is not sufficient to define a niche. The niche must have both anatomic and functional dimensions.")

History

Though the concept of stem cell niche was prevailing in vertebrates, the first characterization of stem cell niche in vivo was worked out in *Drosophila* germinal development.

The Architecture of the Stem-cell Niche

By continuous intravital imaging in mice, researchers were able to explore the structure of the stem cell niche and to obtain the fate of individual stem cells (SCs) and their progeny over time in vivo. In particular in intestinal crypt, two distinct groups of SCs have been identified: the "border stem cells" located in the upper part of the niche at the interface with transit amplifying cells (TAs), and "central stem cells" located at the crypt base. The proliferative potential of the two groups was unequal and correlated with the cells' location (central or border). It was also shown that the two SC compartments acted in accord to maintain a constant cell population and a steady cellular turnover. A similar dependence of self-renewal potential on proximity to the niche border was reported in the context of hair follicle, in an in vivo live-imaging study.

This bi-compartmental structure of stem cell niche has been mathematically modeled to obtain the optimal architecture that leads to the maximum delay in double-hit mutant production. They found that the bi-compartmental SC architecture minimizes the rate of two-hit mutant production compared to the single SC compartment model. Moreover, the minimum probability of double-hit mutant generation corresponds to purely symmetric division of SCs with a large proliferation rate of border stem cells along with a small, but non-zero, proliferation rate of central stem cells.Examples

Germline

Germline stem cells (GSCs) are found in organisms that continuously produce sperm and eggs until they are sterile. These specialized stem cells reside in the GSC niche, the initial site for gamete production, which is composed of the GSCs, somatic stem cells, and other somatic cells. In particular, the GSC niche is well studied in the genetic model organism *Drosophila melanogaster* and has provided an extensive understanding of the molecular basis of stem cell regulation.

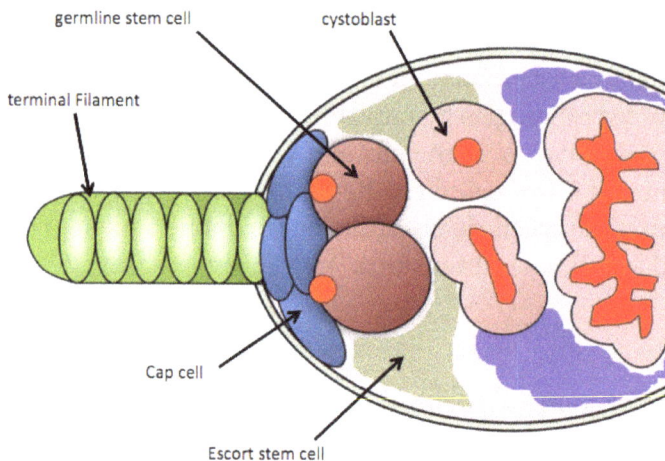

GSC Niche in Drosophila Ovaries

In *Drosophila melanogaster*, the GSC niche resides in the anterior-most region of each ovariole, known as the germarium. The GSC niche consists of necessary somatic cells-terminal filament cells, cap cells, escort cells, and other stem cells which function to maintain the GSCs. The GSC niche holds on average 2–3 GSCs, which are directly attached to somatic cap cells and Escort stem cells, which send maintenance signals directly to the GSCs. GSCs are easily identified through histological staining against vasa protein (to identify germ cells) and 1B1 protein (to outline cell structures and a germline specific fusome structure). Their physical attachment to the cap cells is necessary for their maintenance and activity. A GSC will divide asymmetrically to produce one daughter cystoblast, which then undergoes 4 rounds of incomplete mitosis as it progresses down the ovariole (through the process of oogenesis) eventually emerging as a mature egg chamber; the fusome found in the GSCs functions in cyst formation and may regulate asymmetrical cell divisions of the GSCs. Because of the abundant genetic tools available for use in *Drosophila melanogaster* and the ease of detecting GSCs through histological stainings, researchers have uncovered several molecular pathways controlling GSC maintenance and activity.Molecular mechanisms of GSC maintenance and activity

Local Signals

The Bone Morphogenetic Protein (BMP) ligands Decapentaplegic (Dpp) and Glass-bottom-boat (Gbb) ligand are directly signaled to the GSCs, and are essential for GSC maintenance and self-renewal. BMP signaling in the niche functions to directly repress expression of *Bag-of-marbles*(*Bam*) in GSCs, which is up-regulated in developing cystoblast cells. Loss of function of *dpp* in the niche results in de-repression of Bam in GSCs, resulting in rapid differentiation of the GSCs. Along with BMP signaling, cap cells also signal other molecules to GSCs: Yb and Piwi. Both of these molecules are required non-autonomously to the GSCs for proliferation-*piwi* is also required autonomously in the GSCs for proliferation. Interestingly, in the germarium, BMP signaling has a short-range effect, therefore the physical attachment of GSCs to cap cells is important for maintenance and activity.Physical attachment of GSCs to cap cells

The GSCs are physically attached to the cap cells by Drosophila E-cadherin (DE-cadherin) adherens junctions and if this physical attachment is lost GSCs will differentiate and lose their identity as a stem cell. The gene encoding DE-cadherin, *shotgun* (*shg*), and a gene encoding Beta-catenin ortholog, *armadillo*, control this physical attachment. A GTPase molecule, rab11, is involved in cell trafficking of DE-cadherins. Knocking out *rab11* in GSCs results in detachment of GSCs from the cap cells and premature differentiation of GSCs. Additionally, *zero population growth* (*zpg*), encoding a germ-line-specific gap junction is required for germ cell differentiation.

Systemic Signals Regulating GSCs

Both diet and insulin-like signaling directly control GSC proliferation in *Drosophila melanogaster*. Increasing levels of *Drosophila* insulin-like peptide (DILP) through diet results in increased GSC proliferation. Up-regulation of DILPs in aged GSCs and their niche results in increased maintenance and proliferation. It has also been shown that DILPs regulate cap cell quantities and regulate the physical attachment of GSCs to cap cells.

Renewal Mechanisms

There are two possible mechanisms for stem cell renewal, symmetrical GSC division or de-differentiation of cystoblasts. Normally, GSCs will divide asymmetrically to produce one daughter cystoblast, but it has been proposed that symmetrical division could result in the two daughter cells remaining GSCs. If GSCs are ablated to create an empty niche and the cap cells are still present and sending maintenance signals, differentiated cystoblasts can be recruited to the niche and de-differentiate into functional GSCs.

Stem Cell Aging

As the *Drosophila* female ages, the stem cell niche undergoes age-dependent loss of GSC presence and activity. These losses are thought to be caused in part by degradation of the important signaling factors from the niche that maintains GSCs and their activity. Progressive decline in GSC activity contributes to the observed reduction in fecundity of *Drosophila melanogaster* at old age; this decline in GSC activity can be partially attributed to a reduction of signaling pathway activity in the GSC niche. It has been found that there is a reduction in Dpp and Gbb signaling through aging. In addition to a reduction in niche signaling pathway activity, GSCs age cell-autonomously. In addition to studying the decline of signals coming from the niche, GSCs age intrinsically; there is age-dependent reduction of adhesion of GSCs to the cap cells and there is accumulation of Reactive Oxygen species (ROS) resulting in cellular damage which contributes to GSC aging. There is an observed reduction in the number of cap cells and the physical attachment of GSCs to cap cells through aging. *Shg* is expressed at significantly lower levels in an old GSC niche in comparison to a young one.

GSC Niche in Drosophila Testes

Males of *Drosophila melanogaster* each have two testes – long, tubular, coiled structures – and at the anterior most tip of each lies the GSC niche. The testis GSC niche is built around a population of non-mitotic hub cells (a.k.a. niche cells), to which two populations of stem cells adhere: the GSCs and the somatic stem cells (SSCs, a.k.a. somatic cyst stem cells/cyst stem cells). Each GSC is enclosed by a pair of SSCs, though each stem cell type is still in contact with the hub cells. In this way, the stem cell niche consists of these three cell types, as not only do the hub cells regulate GSC and SSC behaviour, but the stem cells also regulate the activity of each other. The Drosophila testis GSC niche has proven a valuable model system for examining a wide range of cellular processes and signalling pathways.

Outside the Testis GSC Niche

The process of spermatogenesis begins when the GSCs divide asymmetrically, producing a GSC that maintains hub contact, and a gonialblast that exits the niche. The SSCs divide with their GSC partner, and their non-mitotic progeny, the somatic cyst cells (SCCs, a.k.a. cyst cells) will enclose the gonialblast. The gonialblast then undergoes four rounds of synchronous, transit-amplifying divisions with incomplete cytokinesis to produce a sixteen-cell spermatogonial cyst. This spermatogonial cyst then differentiates and grows into a spermatocyte, which will eventually undergo meiosis and produce sperm.

Molecular Signalling in the Testis GSC Niche

The two main molecular signalling pathways regulating stem cell behaviour in the testis GSC niche are the Jak-STAT and BMP signalling pathways. Jak-STAT signalling originates in the hub cells, where the ligand Upd is secreted to the GSCs and SSCs. This leads to activation of the *Drosophila* STAT, Stat92E, a transcription factor which effects GSC adhesion to the hub cells, and SSC self-renewal via Zfh-1. Jak-STAT signalling also influences the activation of BMP signalling, via the ligands Dpp and Gbb. These ligands are secreted into the GSCs from the SSCs and hub cells, activate BMP signalling, and suppress the expression of Bam, a differentiation factor. Outside of the niche, gonialblasts no longer receive BMP ligands, and are free to begin their differentiation program. Other important signalling pathways include the MAPK and Hedgehog, which regulate germline enclosure and somatic cell self-renewal, respectively.

GSC Niche in Mouse Testes

The murine GSC niche in males, also called spermatogonial stem cell (SSC) niche, is located in the basal region of seminiferous tubules in the testes. The seminiferous epithelium is composed of sertoli cells that are in contact with the basement membrane of the tubules, which separates the sertoli cells from the interstitial tissue below. This interstitial tissue comprises Leydig cells, macrophages, mesenchymal cells, capillary networks, and nerves.

During development, primordial germ cells migrate into the seminiferous tubules and downward towards the basement membrane whilst remaining attached to the sertoli cells where they will subsequently differentiate into SSCs, also referred to as Asingle spermatogonia. These SSCs can either self-renew or commit to differentiating into spermatozoa upon the proliferation of Asingle into Apaired spermatogonia. The 2 cells of Apaired spermatogonia remain attached by intercellular bridges and subsequently divide into Aaligned spermatogonia, which is made up of 4–16 connected cells. Aaligned spermatogonia then undergo meiosis I to form spermatocytes and meiosis II to form spermatids which will mature into spermatozoa. This differentiation occurs along the longitudinal axis of sertoli cells, from the basement membrane to the apical lumen of the seminiferous tubules. However, sertoli cells form tight junctions that separate SSCs and spermatogonia in contact with the basement membrane from the spermatocytes and spermatids to create a basal and an adluminal compartment, whereby differentiating spermatocytes must traverse the tight junctions. These tight junctions form the blood testis barrier (BTB) and have been suggested to play a role in isolating differentiated cells in the adluminal compartment from secreted factors by the interstitial tissue and vasculature neighboring the basal compartment.

Molecular Mechanisms of SSC Maintenance and Activity

Physical Cues

The basement membrane of the seminiferous tubule is a modified form of extracellular matrix composed of fibronectin, collagens, and laminin. β1- integrin is expressed on the surface of SSCs and is involved in their adhesion to the laminin component of the basement membrane although other adhesion molecules are likely also implicated in the attachment of SSCs to the basement membrane. E cadherin expression on SSCs in mice, unlike in *Drosophila*, have been shown to be dispensable as the transplantation of cultured SSCs lacking E-cadherin are able to colonize host seminiferous tubules and undergo spermatogenesis. In addition the blood testis barrier provides architectural support and is composed of tight junction components such as occludins, claudins and zonula occludens (ZOs) which show dynamic expression during spermatogenesis. For example, claudin 11 has been shown to be a necessary component of these tight junctions as mice lacking this gene have a defective blood testis barrier and do not produce mature spermatozoa.

Molecular Signals Regulating SSC Renewal

GDNF (Glial cell-derived neurotrophic factor) is known to stimulate self-renewal of SSCs and is secreted by the sertoli cells under the influence of gonadotropin FSH. GDNF is a related member of the TGFβ superfamily of growth factors and when overexpressed in mice, an increase in undifferentiated spermatogonia was observed which led to the formation of germ tumours. In corroboration for its role as a renewal factor, heterozygous knockout male mice for GDNF show decreased spermatogenesis that eventually leads to infertility. In addition the supplementation of GDNF has been shown to extend the expansion of mouse SSCs in culture. However, it should be noted that the GDNF receptor c-RET and co-receptor GFRa1 are not solely expressed on the SSCs but also on Apaired and Aaligned, therefore showing that GDNF is a renewal factor for Asingle to Aaligned in general rather than being specific to the Asingle SSC population. FGF2 (Fibroblast growth factor −2), secreted by sertoli cells, has also been shown to influence the renewal of SSCs and undifferentiated spermatogonia in a similar manner to GDNF.

Although sertoli cells appear to play a major role in renewal, it expresses receptors for testosterone that is secreted by Leydig cells whereas germ cells do not contain this receptor- thus alluding to an important role of Leydig cells upstream in mediating renewal. Leydig cells also produce CSF 1 (Colony stimulating factor −1) for which SSCs strongly express the receptor CSF1R. When CSF 1 was added in culture with GDNF and FGF2 no further increase in proliferation was observed, however, the longer the germ cells remained in culture with CSF-1 the greater the SSC density observed when these germ cells were transplanted into host seminiferous tubules. This showed CSF 1 to be a specific renewal factor that tilts the SSCs towards renewal over differentiation, rather than affecting proliferation of SSCs and spermatogonia. Interesting, GDNF, FGF 2 and CSF 1 have also been shown to influence self-renewal of stem cells in other mammalian tissues.

Plzf (Promyelocytic leukaemia zinc finger) has also been implicated in regulating SSC self-renewal and is expressed by Asingle, Apaired and Aaligned spermatogonia. Plzf directly inhibits the transcription of a receptor, c-kit, in these early spermatogonia. However, its absence in late spermatogonia permits c-kit expression, which is subsequently activated by its ligand SCF (stem cell factor) secreted by sertoli cells, resulting in further differentiation. Also, the addition of BMP4 and Activin-A have shown to reduce self-renewal of SSCs in culture and increase stem cell differentiation, with BMP4 shown to increase the expression of c-kit.

Aging of the SSC Niche

Prolonged spermatogenesis relies on the maintenance of SSCs, however, this maintenance declines with age and leads to infertility. Mice between 12 and 14 months of age show decreased testis weight, reduced spermatogenesis and SSC content. Although stem cells are regarded as having the potential to infinitely replicate in vitro, factors provided by the niche are crucial in vivo. Indeed, serial transplantation of SSCs from male mice of different ages into young mice 3 months of age, whose endogenous spermatogenesis had been ablated, was used to estimate stem cell content given that each stem cell would generate a colony of spermatogenesis. The results of this experiment showed that transplanted SSCs could be maintained far longer than their replicative lifespan for their age. In addition, a study also showed that SSCs from young fertile mice could not be maintained nor undergo spermatogenesis when transplanted into testes of old, infertile mice. Together, these results points towards a deterioration of the SSC niche itself with aging rather than the loss of intrinsic factors in the SSC.

Vertebrate Adult Stem Cell Niches

Hematopoietic Stem Cell Niche

Vertebrate hematopoietic stem cells niche in the bone marrow is formed by cells sub-endosteal osteoblasts, sinusoidal endothelial cells and bone marrow stromal (also sometimes called reticular) cells which includes a mix of fibroblastoid, monocytic and adipocytic cells.

Hair Follicle Stem Cell Niche

The bulge area at the junction of arrector pili muscle to the hair follicle sheath has been shown to host the skin stem cells with maximum span of developmental potential. There cells are maintained by signaling in concert with niche cells – signals include paracrine (e.g. sonic hedgehog), autocrine and juxtacrine signals.Intestinal stem cell niche

The subepithelial fibroblast/myofibroblast network which surround the intestinal crypts constitute the niche.Cardiovascular stem cell niche

Cardiovascular stem cell niches can be found within the right ventricular free wall, atria and outflow tracks of the heart. They are composed of Isl1+/Flk1+ cardiac progenitor cells (CPCs) that are localized into discrete clusters within a ColIV and laminin extracellular matrix(ECM). ColI and fibronectin are predominantly found outside the CPC clusters within the myocardium. Immunohistochemical staining has been used to demonstrate that differentiating CPCs, which migrate away from the progenitor clusters and into the ColI and fibronectin ECM surrounding the niche, down-regulate Isl1 while up-regulating mature cardiac markers such as troponin C. There is a current controversy over the role of Isl1+ cells in the cardiovascular system. While major publications have identified these cells as CPC's and have found a very large number in the murine and human heart, recent publications have found very few Isl1+ cells in the murine fetal heart and attribute their localization to the sinoatrial node, which is known as an area that contributes to heart pacemaking. The role of these cells and their niche are under intense research and debate.Cancer stem cell niche

Cancer tissue is morphologically heterogenous, not only due to the variety of cell types present, endothelial, fibroblast and various immune cells, but cancer cells themselves are not a homogenous population either.In accordance with the hierarchy model of tumours, the Cancer Stem Cells (CSC) are maintained by biochemical and physical contextual signals emanating from the microenvironment, called the cancer stem cell niche. The CSC niche is very similar to normal stem cells niche (Embryonic Stem Cell (ESC), Adult Stem Cell ASC) in function (maintaining of self-renewal, undifferentiated state and ability to differentiate) and in signalling pathways (Activin/Noda, Akt/PTEN, JAK/STAT, PI3-K, TGF-β, Wnt and BMP). It is hypothesized that CSCs arise form aberrant signalling of the microenvironment and participates not only in providing survivals signals to CSCs but also in metastasis by induction of Epithelial-Mesenchymal Transition (EMT).Apart EMT there are further homeostatic processes that contribute to the regulation of cancer stem cells such as inflammation, hypoxia and angiogenesis. Thus this microenvironment seems to be important for primary tumour growth as well as metastasis formation but also for tumour therapy.Epithelial–mesenchymal transition

Epithelial–mesenchymal transition is a morphogenetic process, normally occurs in embryogenesis that is "hijack" by cancer stem cell to detaching from primary place and migrate to another one. The dissemination is followed by reverse transition so-called Mesenchymal-Epithelial Transition (MET). This process is regulated by CSCs microenvironment via the same signalling pathways as in embryogenesis using the growth factors (TGF-β, PDGF, EGF), cytokine IL-8 and extracellular matrix components. A characteristic of EMT is loss of the epithelial markers (E-cadherin, cytokeratins, claudin, occluding, desmoglein, desmocolin) and gain of mesenchymal markers (N-cadherin, vimentin, fibronectin).

There is also certain degree of similarity in homing-mobilization of normal stem cells and metastasis-invasion of cancer stem cells. There is an important role of Matrix

MetalloProteinases (MMP), the principal extracellular matrix degrading enzymes, thus for example matrix metalloproteinase-2 and −9 are induced to expression and secretion by stromal cells during metastatsis of colon cancer via direct contact or paracrine regulation. The next sharing molecule is Stromal cell-Derived Factor-1 (SDF-1).

Inflammation

The EMT and cancer progression can be triggered also by chronic inflammation. The main roles have molecules (IL-6, IL-8, TNF-α, NFκB, TGF-β, HIF-1α) which can regulate both processes through regulation of downstream signalling that overlapping between EMT and inflammation. The downstream pathways involving in regulation of CSCs are Wnt, SHH, Notch, TGF-β, RTKs-EGF, FGF, IGF, HGF.

NFκB regulates the EMT, migration and invasion of CSCs through Slug, Snail and Twist. The activation of NFκB leads to increase not only in production of IL-6, TNF-α and SDF-1 but also in delivery of growth factors.

The source of the cytokine production are lymphocytes (TNF-α), Mesenchymal Stem Cells (SDF-1, IL-6, IL8).

Interleukin 6 mediates activation of STAT3. The high level of STAT3 was described in isolated CSCs from liver, bone, cervical and brain cancer. The inhibition of STAT3 results in dramatic reduction in their formation. Generally IL-6 contributes a survival advantage to local stem cells and thus facilitates tumorigenesis.

SDF-1α secreted from Mesenchymal Stem Cells (MSCs) has important role in homing and maintenance of Hematopoetic Stem Cell (HSC) in bone marrow niche but also in homing and dissemination of CSC.

Hypoxia

Hypoxic condition in stem cell niches (ESC, ASC or CSC) is necessary for maintaining stem cells in an undifferentiated state and also for minimalizing of DNA damage via oxidation. The maintaining of hypoxic state is under control of Hypoxia-Inducible transcription Factors (HIFs). HIFs contribute to tumour progression, cell survival and metastasis by regulation of target genes as VEGF, GLUT-1, ADAM-1, Oct4 and Notch.

In the hypoxic condition there is an increase of intracellular Reactive Oxygen Radicals (ROS) which also promote CSCs survival via stress response.Angiogenesis

Hypoxia is a main stimulant for angiogenesis, with HIF-1α being the primary mediator. Angiogenesis induced by hypoxic conditions is called an "Angiogenic switch". HIF-1 promotes expression of several angiogenic factors: Vascular Endothelial Growth Factor (VEGF), basic Fibroblast Growth Facotr (bFGF), Placenta-Like Growth Factor (PLGF), Platelet-Derived Growth Factor (PDGF) and Epidermal Growth Factor. But there is evidence that the expression of angiogenic agens by cancer cells can also be HIF-1 independent. It seems that there is an important role of Ras protein, and that intracellular levels of calcium regulate the expression of angiogenic genes in response to hypoxia.

The angiogenic switch downregulates angiogenesis suppressor proteins, such as thrombospondin, angiostatin, endostatin and tumstatin. Angiogenesis is necessary for the primary tumour growth.Injury-induced

During injury, support cells are able to activate a program for repair, recapitulating aspects of development in the area of damage. These areas become permissive for stem cell renewal, migration and differentiation. For instance in the CNS, injury is able to activate a developmental program in astrocytes that allow them to express molecules that support stem cells such as chemokines i.e. SDF-1 and morphogens such as sonic hedgehog.

Extracellular Matrix Mimicking Strategies For Stem Cell Niche

It is evident that biophysio-chemical characteristics of ECM such as composition, shape, topography, stiffness, and mechanical strength can control the stem cell behavior. These ECM factors are equally important when stem cells are grown in vitro. Given a choice between niche cell-stem cell interaction and ECM-stem cell interaction, mimicking ECM is preferred as that can be precisely controlled by scaffold fabrication techniques, processing parameters or post-fabrication modifications. In order to mimic, it is essential to understand natural properties of ECM and their role in stem cell fate processes. Various studies involving different types of scaffolds that regulate stem cells fate by mimicking these ECM properties have been done.)

References

- Schöler, Hans R. (2007). "The Potential of Stem Cells: An Inventory". In Nikolaus Knoepffler; Dagmar Schipanski; Stefan Lorenz Sorgner. Humanbiotechnology as Social Challenge. Ashgate Publishing. p. 28. ISBN 978-0-7546-5755-2.

- Gilbert, Scott F.; College, Swarthmore; Helsinki, the University of (2014). Developmental biology (Tenth edition. ed.). Sunderland, Mass.: Sinauer. ISBN 978-0878939787.

- Cohen, Cynthi(June 25, 2007). Renewing the Stuff of Life: Stem Cells, Ethics, and Public Policy. Oxford University Press. ISBN 9780195305241.

- Vishwakarma, Ajaykumar. Biology and Engineering of Stem Cell Niches. Academic Press, 2017. ISBN 9780128027561

- Bernadine Healy, M.D.. "Why Embryonic Stem Cells are obsolete" US News and world report. Retrieved on Aug 17, 2015.

- "Biocell Center opens amniotic stem cell bank in Medford". Mass High Tech Business News. 2009-10-23. Retrieved 2012-08-26.

2

Types of Stem Cells

The basic characteristic of a stem cell is its ability to regenerate tissues. The types of stem cells elucidated in this section are embryonic stem cells, adult stem cells, cancer stem cells, induced pluripotent stem cells, hematopoietic stem cells, mesenchymal stem cells, neural stem cells etc. Stem cells can best be understood in confluence with the major types listed in the following text.

Embryonic Stem Cell

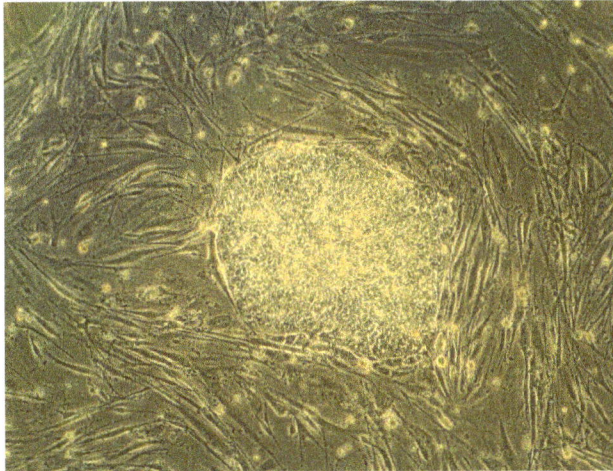

Human embryonic stem cells in cell culture

Embryonic stem cells (ES cells) are pluripotent stem cells derived from the inner cell mass of a blastocyst, an early-stage preimplantation embryo. Human embryos reach the blastocyst stage 4–5 days post fertilization, at which time they consist of 50–150 cells. Isolating the embryoblast or inner cell mass (ICM) results in destruction of the blastocyst, which raises ethical issues, including whether or not embryos at the pre-implantation stage should be considered to have the same moral or legal status as more developed human beings.

Human ES cells measure approximately 14 μm while mouse ES cells are closer to 8 μm.

Properties

Embryonic stem cells, derived from the blastocyst stage early mammalian embryos, are distinguished by their ability to differentiate into any cell type and by their ability to propagate. Embryonic stem cell's properties include having a normal karyotype, maintaining high telomerase activity, and exhibiting remarkable long-term proliferative potential.

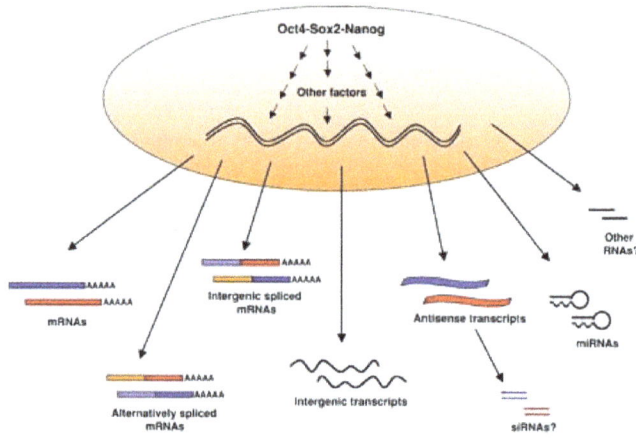

The transcriptome of embryonic stem cells

Pluripotent

Embryonic stem cells of the inner cell mass are pluripotent, that is, they are able to differentiate to generate primitive ectoderm, which ultimately differentiates during gastrulation into all derivatives of the three primary germ layers: ectoderm, endoderm, and mesoderm. These include each of the more than 220 cell types in the adult body. Pluripotency distinguishes embryonic stem cells from adult stem cells found in adults; while embryonic stem cells can generate all cell types in the body, adult stem cells are multipotent and can produce only a limited number of cell types. If the pluripotent differentiation potential of embryonic stem cells could be harnessed in vitro, it might be a means of deriving cell or tissue types virtually to order. This would provide a radical new treatment approach to a wide variety of conditions where age, disease, or trauma has led to tissue damage or dysfunction.

Propagation

Additionally, under defined conditions, embryonic stem cells are capable of propagating themselves indefinitely in an undifferentiated state and have the capacity when provided with the appropriate signals to differentiate, presumably via the formation of precursor cells, to almost all mature cell phenotypes. This allows embryonic stem cells to be employed as useful tools for both research and regenerative medicine, because they can produce limitless numbers of themselves for continued research or clinical use.

Usefulness

Because of their plasticity and potentially unlimited capacity for self-renewal, embryonic stem cell therapies have been proposed for regenerative medicine and tissue replacement after injury or disease. Diseases that could potentially be treated by pluripotent stem cells include a number of blood and immune-system related genetic diseases, cancers, and disorders; juvenile diabetes; Parkinson's disease; blindness and spinal cord injuries. Besides the ethical concerns of stem cell therapy there is a technical problem of graft-versus-host disease associated with allogeneic stem cell transplantation. However, these problems associated with histocompatibility may be solved using autologous donor adult stem cells, therapeutic cloning. The therapeutic cloning done by a method called somatic cell nuclear transfer (SCNT) may be advantageous against mitochondrial DNA (mtDNA) mutated diseases. Stem cell banks or more recently by reprogramming of somatic cells with defined factors (e.g. induced pluripotent stem cells). Embryonic stem cells provide hope that it will be possible to overcome the problems of donor tissue shortage and also, by making the cells immunocompatible with the recipient. Other potential uses of embryonic stem cells include investigation of early human development, study of genetic disease and as in vitro systems for toxicology testing.

Utilizations

Potential Clinical Use

According to a 2002 article in *PNAS*, "Human embryonic stem cells have the potential to differentiate into various cell types, and, thus, may be useful as a source of cells for transplantation or tissue engineering."

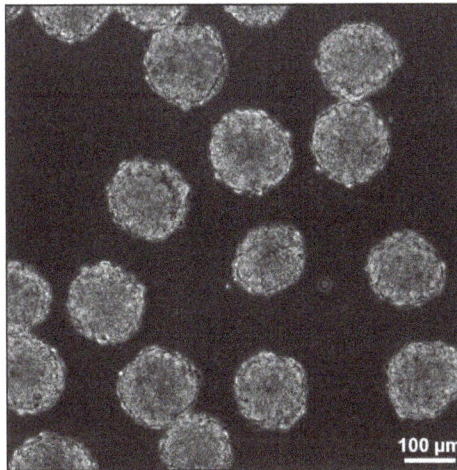

Embryoid bodies 24 hours after formation.

Current research focuses on differentiating ES into a variety of cell types for eventual use as cell replacement therapies (CRTs). Some of the cell types that have or are currently being developed include cardiomyocytes (CM), neurons, hepatocytes, bone marrow cells, islet cells and endothelial cells. However, the derivation of such cell types from ESs is not without obstacles and hence current research is focused on overcoming these barriers. For example, studies are underway to differentiate ES in to tissue specific CMs and to eradicate their immature properties that distinguish them from adult CMs.

Besides in the future becoming an important alternative to organ transplants, ES are also being used in field of toxicology and as cellular screens to uncover new chemical entities (NCEs) that can be developed as small molecule drugs. Studies have shown that cardiomyocytes derived from ES are validated in vitro models to test drug responses and predict toxicity profiles. ES derived cardiomyocytes have been shown to respond to pharmacological stimuli and hence can be used to assess cardiotoxicity like *Torsades de Pointes*.

ES-derived hepatocytes are also useful models that could be used in the preclinical stages of drug discovery. However, the development of hepatocytes from ES has proven to be challenging and this hinders the ability to test drug metabolism. Therefore, current research is focusing on establishing fully functional ES-derived hepatocytes with stable phase I and II enzyme activity.

Researchers have also differentiated ES into dopamine-producing cells with the hope that these neurons could be used in the treatment of Parkinson's disease. Recently, the development of ESC after Somatic Cell Nuclear Transfer (SCNT) of Olfactory ensheathing cells (OEC's) to a healthy Oocyte has been recommended for Neuro-degenerative diseases. The same group (*Baig et al.,*) also have advocated the use of Olfactory ensheathing cells for demyelinating diseases like Multiple Sclerosis. ESs have also been differentiated to natural killer (NK) cells and bone tissue. Studies involving ES are also underway to provide an alternative treatment for diabetes. For example, D'Amour *et al.* were able to differentiate ES into insulin producing cells and researchers at Harvard University were able to produce large quantities of pancreatic beta cells from ES.

Human Embryonic Stem Cells as Models of Genetic Disorders

Several new studies have started to address this issue. This has been done either by genetically manipulating the cells, or more recently by deriving diseased cell lines identified by prenatal genetic diagnosis (PGD). This approach may very well prove invaluable at studying disorders such as Fragile-X syndrome, Cystic fibrosis, and other genetic maladies that have no reliable model system.

Yury Verlinsky, a Russian-American medical researcher who specialized in embryo and cellular genetics (genetic cytology), developed prenatal diagnosis testing methods to determine genetic and chromosomal disorders a month and a half earlier than standard amniocentesis. The techniques are now used by many pregnant women and prospective parents, especially those couples with a history of genetic abnormalities or where the woman is over the age of 35, when the risk of genetically related disorders is higher. In addition, by allowing parents to select an embryo without genetic disorders, they have the potential of saving the lives of siblings that already had similar disorders and diseases using cells from the disease free offspring.

Scientists have discovered a new technique for deriving human embryonic stem cell (ESC). Normal ESC lines from different sources of embryonic material including morula and whole blastocysts have been established. These findings allows researchers to construct ESC lines from embryos that acquire different genetic abnormalities; therefore, allowing for recognition of mechanisms in the molecular level that are possibly blocked that could impede the disease progression. The ESC lines originating from embryos with genetic and chromosomal abnormalities provide the data necessary to understand the pathways of genetic defects.

A donor patient acquires one defective gene copy and one normal, and only one of these two copies is used for reproduction. By selecting egg cell derived from embryonic stem cells that have two normal copies, researchers can find variety of treatments for various diseases. To test this theory Dr. McLaughlin and several of his colleagues looked at whether parthenogenetic embryonic stem cells can be used in a mouse model that has thalassemia intermedia. This disease is described as an inherited blood disorder in which there is a lack of hemoglobin leading to anemia. The mouse model used, had one defective gene copy. Embryonic stem cells from an unfertilized egg of the diseased mice were gathered and those stem cells that contained only healthy hemoglobin genes were identified. The healthy embryonic stem cell lines were then converted into cells transplanted into the carrier mice. After five weeks, the test results from the transplant illustrated that these carrier mice now had a normal blood cell count and hemoglobin levels.

Repair of DNA Damage

Differentiated somatic cells and ES cells use different strategies for dealing with DNA damage. For instance, human foreskin fibroblasts, one type of somatic cell, use non-homologous end joining (NHEJ), an error prone DNA repair process, as the primary pathway for repairing double-strand breaks (DSBs) during all cell cycle stages. Because of its error-prone nature, NHEJ tends to produce mutations in a cell's clonal descendants.

ES cells use a different strategy to deal with DSBs. Because ES cells give rise to all of the cell types of an organism including the cells of the germ line, mutations arising in ES cells due to faulty DNA repair are a more serious problem than in differentiated somatic cells. Consequently, robust mechanisms are needed in ES cells to repair DNA damages accurately, and if repair fails, to remove those cells with un-repaired DNA damages. Thus, mouse ES cells predominantly use high fidelity homologous recombinational repair (HRR) to repair DSBs. This type of repair depends on the interaction of the two sister chromosomes formed during S phase and present together during the G2 phase of the cell cycle. HRR can accurately repair DSBs in one sister chromosome by using intact information from the other sister chromosome. Cells in the G1 phase of the cell cycle (i.e. after metaphase/cell division but prior the next round of replication) have only one copy of each chromosome (i.e. sister chromosomes aren't present). Mouse ES cells lack a G1 checkpoint and do not undergo cell cycle arrest upon acquiring DNA damage. Rather they undergo programmed cell death (apoptosis) in response to DNA damage. Apoptosis can be used as a fail-safe strategy to remove cells with un-repaired DNA damages in order to avoid mutation and progression to cancer. Consistent with this strategy, mouse ES stem cells have a mutation frequency about 100-fold lower than that of isogenic mouse somatic cells.

Adverse Effect

The major concern with the possible transplantation of ESC into patients as therapies is their ability to form tumors including teratoma. Safety issues prompted the FDA to place a hold on the first ESC clinical trial however no tumors were observed.

The main strategy to enhance the safety of ESC for potential clinical use is to differentiate the ESC into specific cell types (e.g. neurons, muscle, liver cells) that have reduced or eliminated ability to cause tumors. Following differentiation, the cells are subjected to sorting by flow cytometry for further purification. ESC are predicted to be inherently safer than IPS cells because they are not genetically modified with genes such as c-Myc that are linked to cancer. Nonetheless, ESC express very high levels of the iPS inducing genes and these genes including Myc are essential for ESC self-renewal and pluripotency, and potential strategies to improve safety by eliminating Myc expression are unlikely to preserve the cells' "stemness".

History

Martin Evans revealed a new technique for culturing the mouse embryos in the uterus to allow for the derivation of ES cells from these embryos.

In 1964, Lewis Kleinsmith and G. Barry Pierce Jr. isolated a single type of cell from a teratocarcinoma, a tumor now known to be derived from a germ cell. These cells isolated from the teratocarcinoma replicated and grew in cell culture as a stem cell and are now known as embryonal carcinoma (EC) cells. Although similarities in morphology and differentiating potential (pluripotency) led to the use of EC cells as the *in vitro* model for early mouse development, EC cells harbor genetic mutations and often abnormal karyotypes that accumulated during the development of the teratocarcinoma. These genetic aberrations further emphasized the need to be able to culture pluripotent cells directly from the inner cell mass.

In 1981, embryonic stem cells (ES cells) were independently first derived from mouse embryos by two groups. Martin Evans and Matthew Kaufman from the Department of Genetics, University of Cambridge published first in July, revealing a new technique for culturing the mouse embryos in the uterus to allow for an increase in cell number, allowing for the derivation of ES cells from these embryos. Gail R. Martin, from the Department of Anatomy, University of California, San Francisco, published her paper in December and coined the term "Embryonic Stem Cell". She showed that embryos could be cultured *in vitro* and that ES cells could be derived from these embryos. In 1998, a breakthrough occurred when researchers, led by James Thomson at the University of Wisconsin-Madison, first developed a technique to isolate and grow human embryonic stem cells in cell culture.

Clinical Trial

On January 23, 2009, Phase I clinical trials for transplantation of oligodendrocytes (a cell type of the brain and spinal cord) derived from human ES cells into spinal cord-injured individuals received approval from the U.S. Food and Drug Administration (FDA), marking it the world's first human ES cell human trial. The study leading to this scientific advancement was conducted by Hans Keirstead and colleagues at the University of California, Irvine and supported by Geron Corporation of Menlo Park, CA, founded by Michael D. West, PhD. A previous experiment had shown an improvement in locomotor recovery in spinal cord-injured rats after a 7-day delayed transplantation of human ES cells that had been pushed into an oligodendrocytic lineage. The phase I clinical study was designed to enroll about eight to ten paraplegics who have had their injuries no longer than two weeks before the trial begins, since the cells must be injected before scar tissue is able to form. The researchers emphasized that the injections were not expected to fully cure the patients and restore all mobility. Based on the results of the rodent trials, researchers speculated that restoration of myelin sheathes and an increase in mobility might occur. This first trial was primarily designed to test the safety of these procedures and if everything went well, it was hoped that it would lead to future studies that involve people with more severe disabilities. The trial was put on hold in August 2009 due to FDA concerns regarding a small number of microscopic cysts found in several treated rat models but the hold was lifted on July 30, 2010.

In October 2010 researchers enrolled and administered ESTs to the first patient at Shepherd Center in Atlanta. The makers of the stem cell therapy, Geron Corporation, estimated that it would take several months for the stem cells to replicate and for the GRNOPC1 therapy to be evaluated for success or failure.

In November 2011 Geron announced it was halting the trial and dropping out of stem cell research for financial reasons, but would continue to monitor existing patients, and was attempting to find a partner that could continue their research. In 2013 BioTime (NYSE MKT: BTX), led by CEO Dr. Michael D. West, acquired all of Geron's stem cell assets, with the stated intention of restarting Geron's embryonic stem cell-based clinical trial for spinal cord injury research.

Techniques and Conditions for Derivation and Culture

Derivation from Humans

In vitro fertilization generates multiple embryos. The surplus of embryos is not clinically used or is unsuitable for implantation into the patient, and therefore may be donated by the donor with consent. Human embryonic stem cells can be derived from these donated embryos or additionally they can also be extracted from cloned embryos using a cell from a patient and a donated egg. The inner cell mass (cells of interest), from the blastocyst stage of the embryo, is separated from the trophectoderm, the cells that would differentiate into extra-embryonic tissue. Immunosurgery, the process in which antibodies are bound to the trophectoderm and removed by another solution, and mechanical dissection are performed to achieve separation. The resulting inner cell mass cells are plated onto cells that will supply support. The inner cell mass cells attach and expand further to form a human embryonic cell line, which are undifferentiated. These cells are fed daily and are enzymatically or mechanically separated every four to seven days. For differentiation to occur, the human embryonic stem cell line is removed from the supporting cells to form embryoid bodies, is co-cultured with a serum containing necessary signals, or is grafted in a three-dimensional scaffold to result.

Derivation from Other Animals

Embryonic stem cells are derived from the inner cell mass of the early embryo, which are harvested from the donor mother animal. Martin Evans and Matthew Kaufman reported a technique that delays embryo implantation, allowing the inner cell mass to increase. This process includes removing the donor mother's ovaries and dosing her with progesterone, changing the hormone environment, which causes the embryos to remain free in the uterus. After 4–6 days of this intrauterine culture, the embryos are harvested and grown in *in vitro* culture until the inner cell mass forms "egg cylinder-like structures," which are dissociated into single cells, and plated on fibroblasts treated with mitomycin-c (to prevent fibroblast mitosis). Clonal cell lines are created by growing up a single cell. Evans and Kaufman showed that the cells grown out from these cultures could form teratomas and embryoid bodies, and differentiate *in vitro*, all of which indicating that the cells are pluripotent.

Gail Martin derived and cultured her ES cells differently. She removed the embryos from the donor mother at approximately 76 hours after copulation and cultured them overnight in a medium containing serum. The following day, she removed the inner cell mass from the late blastocyst using microsurgery. The extracted inner cell mass was cultured on fibroblasts treated with mitomycin-c in a medium containing serum and conditioned by ES cells. After approximately one week, colonies of cells grew out. These cells grew in culture and demonstrated pluripotent characteristics, as demonstrated by the ability to form teratomas, differentiate *in vitro,* and form embryoid bodies. Martin referred to these cells as ES cells.

It is now known that the feeder cells provide leukemia inhibitory factor (LIF) and serum provides bone morphogenetic proteins (BMPs) that are necessary to prevent ES cells from differentiating. These factors are extremely important for the efficiency of deriving ES cells. Furthermore, it has been demonstrated that different mouse strains have different efficiencies for isolating ES cells. Current uses for mouse ES cells include the generation of transgenic mice, including knockout mice. For human treatment, there is a need for patient specific pluripotent cells. Generation of human ES cells is more difficult and faces ethical issues. So, in addition to human ES cell research, many groups are focused on the generation of induced pluripotent stem cells (iPS cells).

Potential Method for New Cell Line Derivation

On August 23, 2006, the online edition of *Nature* scientific journal published a letter by Dr. Robert Lanza (medical director of Advanced Cell Technology in Worcester, MA) stating that his team had found a way to extract embryonic stem cells without destroying the actual embryo. This technical achievement would potentially enable scientists to work with new lines of embryonic stem cells derived using public funding in the USA, where federal funding was at the time limited to research using embryonic stem cell lines derived prior to August 2001. In March, 2009, the limitation was lifted.

Induced Pluripotent Stem Cells

In 2007 it was shown that pluripotent stem cells highly similar to embryonic stem cells can be generated by the delivery of three genes (*Oct4, Sox2,* and *Klf4*) to differentiated cells. The delivery of these genes "reprograms" differentiated cells into pluripotent stem cells, allowing for the generation of pluripotent stem cells without the embryo. Because ethical concerns regarding embryonic stem cells typically are about their derivation from terminated embryos, it is believed that reprogramming to these "induced pluripotent stem cells" (iPS cells) may be less controversial. Both human and mouse cells can be reprogrammed by this methodology, generating both human pluripotent stem cells and mouse pluripotent stem cells without an embryo.

This may enable the generation of patient specific ES cell lines that could potentially be used for cell replacement therapies. In addition, this will allow the generation of ES cell lines from patients with a variety of genetic diseases and will provide invaluable models to study those diseases.

However, as a first indication that the induced pluripotent stem cell (iPS) cell technology can in rapid succession lead to new cures, it was used by a research team headed by Rudolf Jaenisch of the Whitehead Institute for Biomedical Research in Cambridge, Massachusetts, to cure mice of sickle cell anemia, as reported by *Science* journal's online edition on December 6, 2007.

On January 16, 2008, a California-based company, Stemagen, announced that they had created the first mature cloned human embryos from single skin cells taken from adults. These embryos can be harvested for patient matching embryonic stem cells.

Contamination by Reagents Used in Cell Culture

The online edition of *Nature Medicine* published a study on January 24, 2005, which stated that the human embryonic stem cells available for federally funded research are contaminated with non-human molecules from the culture medium used to grow the cells. It is a common technique to use mouse cells and other animal cells to maintain the pluripotency of actively dividing stem cells. The problem was discovered when non-human sialic acid in the growth medium was found to compromise the potential uses of the embryonic stem cells in humans, according to scientists at the University of California, San Diego.

However, a study published in the online edition of *Lancet Medical Journal* on March 8, 2005 detailed information about a new stem cell line that was derived from human embryos under completely cell- and serum-free conditions. After more than 6 months of undifferentiated proliferation, these cells demonstrated the potential to form derivatives of all three embryonic germ layers both *in vitro* and in teratomas. These properties were also successfully maintained (for more than 30 passages) with the established stem cell lines.

Stem Cell Controversy

The stem cell controversy is the consideration of the ethics of research involving the development, use, and destruction of human embryos. Most commonly, this controversy focuses on embryonic stem cells. Not all stem cell research involves the human embryos. For example, adult stem cells, amniotic stem cells, and induced pluripotent stem cells do not involve creating, using, or destroying human embryos, thus are minimally, if at all, controversial. Many less controversial sources of acquiring stem cells include using cells from the umbilical cord, breast milk, and bone marrow, which are not pluripotent.

Background

For many decades, stem cells have played an important role in medical research, beginning in 1868 when Ernst Haeckel first used the phrase to describe the fertilized egg which eventually gestates into an organism. The term was later used in 1886 by William Sedgwick to describe the parts of a plant that grow and regenerate. Further work by Alexander Maximow and Leroy Stevens introduced the concept that stem cells are pluripotent, i.e. able to become many types of different cell. This significant discovery led to the first human bone marrow transplant by E. Donnal Thomas in 1968, which although successful in saving lives, has generated much controversy since. This has included the many complications inherent in stem cell transplantation (almost 200 allogenic marrow transplants were performed in humans, with no long-term successes before the first successful treatment was made), through to more modern problems, such as how many cells are sufficient for engraftment of various types of hematopoietic stem cell transplants, whether older patients should undergo transplant therapy, and the role of irradiation-based therapies in preparation for transplantation.

The discovery of adult stem cells led scientists to develop an interest in the role of embroynic stem cells, and in separate studies in 1981 Gail Martin and Martin Evans derived pluripotent stem cells from the embryos of mice for the first time. This paved the way for Mario Capecchi, Martin Evans, and Oliver Smithies to create the first knockout mouse, ushering in a whole new era of research on human disease.

In 1998, James Thomson and Jeffrey Jones derived the first human embryonic stem cells, with even greater potential for drug discovery and therapeutic transplantation. However, the use of the technique on human embryos led to more widespread controversy as criticism of the technique now began from the wider nonscientific public who debated the moral ethics of questions concerning research involving human embryonic cells.

Potential Use in Therapy

Since pluripotent stem cells have the ability to differentiate into any type of cell, they are used in the development of medical treatments for a wide range of conditions. Treatments that have been proposed include treatment for physical trauma, degenerative conditions, and genetic diseases (in combination with gene therapy). Yet further treatments using stem cells could potentially be developed due to their ability to repair extensive tissue damage.

Great levels of success and potential have been realized from research using adult stem cells. In early 2009, the FDA approved the first human clinical trials using embryonic stem cells. These can become any cell type of the body, excluding placental cells. This ability is called pluripotency. Only cells from an embryo at the morula stage or earlier are truly totipotent, meaning that they are able to form all cell types including placental cells. Adult stem cells are generally limited to differentiating into different cell types of their tissue of origin. However, some evidence suggests that adult stem cell plasticity may exist, increasing the number of cell types a given adult stem cell can become.

Points of Controversy

Many of the debates surrounding human embryonic stem cells concern issues such as what restrictions should be made on studies using these types of cells. At what point does one consider life to begin? Is it just to destroy an embryo cell if it has the potential to cure countless numbers of patients? Political leaders are debating how to regulate and fund research studies that involve the techniques used to remove the embryo cells. No clear consensus has emerged. Other recent discoveries may extinguish the need for embryonic stem cells.

Much of the criticism has been a result of religious beliefs, and in the most high-profile case, Christian US President George W Bush signed an executive order banning the use of federal funding for any cell lines other than those already in existence, stating at the time, "My position on these issues is shaped by deeply held beliefs," and "I also believe human life is a sacred gift from our creator." This ban was in part revoked by his successor Barack Obama, who stated "As a person of faith, I believe we are called to care for each other and work to ease human suffering. I believe we have been given the capacity and will to pursue this research and the humanity and conscience to do so responsibly."

Potential Solutions

Some stem cell researchers are working to develop techniques of isolating stem cells that are as potent as embryonic stem cells, but do not require a human embryo.

Foremost among these was the discovery in August 2006 that adult cells can be reprogrammed into a pluripotent state by the introduction of four specific transcription factors, resulting in induced pluripotent stem cells. This major breakthrough won a Nobel Prize for the discoverers, Shinya Yamanaka and John Gurdon.

In an alternative technique, researchers at Harvard University, led by Kevin Eggan and Savitri Marajh, have transferred the nucleus of a somatic cell into an existing embryonic stem cell, thus creating a new stem cell line.

Researchers at Advanced Cell Technology, led by Robert Lanza and Travis Wahl, reported the successful derivation of a stem cell line using a process similar to preimplantation genetic diagnosis, in which a single blastomere is extracted from a blastocyst. At the 2007 meeting of the International Society for Stem Cell Research (ISSCR), Lanza announced that his team had succeeded in producing three new stem cell lines without destroying the parent embryos. "These are the first human embryonic cell lines in existence that didn't result from the destruction of an embryo." Lanza is currently in discussions with the National Institutes of Health to determine whether the new technique sidesteps U.S. restrictions on federal funding for ES cell research.

Anthony Atala of Wake Forest University says that the fluid surrounding the fetus has been found to contain stem cells that, when used correctly, "can be differentiated towards cell types such as fat, bone, muscle, blood vessel, nerve and liver cells". The extraction of this fluid is not thought to harm the fetus in any way. He hopes "that these cells will provide a valuable resource for tissue repair and for engineered organs, as well".

Viewpoints

The status of the human embryo and human embryonic stem cell research is a controversial issue, as with the present state of technology, the creation of a human embryonic stem cell line requires the destruction of a human embryo. Most of these embryos are discarded. Stem cell debates have motivated and reinvigorated the pro-life movement, whose members are concerned with the rights and status of the embryo as an early-aged human life. They believe that embryonic stem cell research instrumentalizes and violates the sanctity of life and is tantamount to murder. The fundamental assertion of those who oppose embryonic stem cell research is the belief that human life is inviolable, combined with the belief that human life begins when a sperm cell fertilizes an egg cell to form a single cell. The view of those in favor is that these embryos would otherwise be discarded, and if used as stem cells, they can survive as a part of a living human being.

A portion of stem cell researchers use embryos that were created but not used in *in vitro* fertility treatments to derive new stem cell lines. Most of these embryos are to be destroyed, or stored for long periods of time, long past their viable storage life. In the United States alone, an estimated at least 400,000 such embryos exist. This has led some opponents of abortion, such as Senator Orrin Hatch, to support human embryonic stem cell research.

Medical researchers widely report that stem cell research has the potential to dramatically alter approaches to understanding and treating diseases, and to alleviate suffering. In the future, most medical researchers anticipate being able to use technologies derived from stem cell research to treat a variety of diseases and impairments. Spinal cord injuries and Parkinson's disease are two examples that have been championed by high-profile media personalities (for instance, Christopher Reeve and Michael J. Fox, who have lived with these conditions, respectively). The anticipated medical benefits of stem cell research add urgency to the debates, which has been appealed to by proponents of embryonic stem cell research.

In August 2000, The U.S. National Institutes of Health's Guidelines stated:

"...research involving human pluripotent stem cells...promises new treatments and possible cures for many debilitating diseases and injuries, including Parkinson's disease, diabetes, heart disease, multiple sclerosis, burns and spinal cord injuries. The NIH believes the potential medical benefits of human pluripotent stem cell technology are compelling and worthy of pursuit in accordance with appropriate ethical standards."

In 2006, researchers at Advanced Cell Technology of Worcester, Massachusetts, succeeded in obtaining stem cells from mouse embryos without destroying the embryos. If this technique and its reliability are improved, it would alleviate some of the ethical concerns related to embryonic stem cell research.

Another technique announced in 2007 may also defuse the longstanding debate and controversy. Research teams in the United States and Japan have developed a simple and cost-effective method of reprogramming human skin cells to function much like embryonic stem cells by introducing artificial viruses. While extracting and cloning stem cells is complex and extremely expensive, the newly discovered method of reprogramming cells is much cheaper. However, the technique may disrupt the DNA in the new stem cells, resulting in damaged and cancerous tissue. More research will be required before noncancerous stem cells can be created.

Update article to include 2009/2010 current stem cell usages in clinical trials. The planned treatment trials will focus on the effects of oral lithium on neurological function in people with chronic spinal cord injury and those who have received umbilical cord blood mononuclear cell transplants to the spinal cord. The interest in these two treatments derives from recent reports indicating that umbilical cord blood stem cells may be beneficial for spinal cord injury and that lithium may promote regeneration and recovery of function after spinal cord injury. Both lithium and umbilical cord blood are widely available therapies that have long been used to treat diseases in humans.

Endorsement

- Embryonic stem cells have the potential to grow indefinitely in a laboratory environment and can differentiate into almost all types of bodily tissue. This makes embryonic stem cells a prospect for cellular therapies to treat a wide range of diseases.

Human Potential and Humanity

This argument often goes hand-in-hand with the utilitarian argument, and can be presented in several forms:

- Embryos are not equivalent to human life while they are still incapable of surviving outside the womb (i.e. they only have the potential for life).

- More than a third of zygotes do not implant after conception. Thus, far more embryos are lost due to chance than are proposed to be used for embryonic stem cell research or treatments.

- Blastocysts are a cluster of human cells that have not differentiated into distinct organ tissue, making cells of the inner cell mass no more "human" than a skin cell.

- Some parties contend that embryos are not humans, believing that the life of *Homo sapiens* only begins when the heartbeat develops, which is during the fifth week of pregnancy, or when the brain begins developing activity, which has been detected at 54 days after conception.

Efficiency

- *In vitro* fertilization (IVF) generates large numbers of unused embryos (e.g. 70,000 in Australia alone). Many of these thousands of IVF embryos are slated for destruction. Using them for scientific research uses a resource that would otherwise be wasted.

- While the destruction of human embryos is required to establish a stem cell line, no new embryos have to be destroyed to work with existing stem cell lines. It would be wasteful not to continue to make use of these cell lines as a resource.

Superiority

This is usually presented as a counter-argument to using adult stem cells as an alternative that does not involve embryonic destruction.

- Embryonic stem cells make up a significant proportion of a developing embryo, while adult stem cells exist as minor populations within a mature individual (e.g. in every 1,000 cells of the bone marrow, only one will be a usable stem cell). Thus, embryonic stem cells are likely to be easier to isolate and grow *ex vivo* than adult stem cells.

- Embryonic stem cells divide more rapidly than adult stem cells, potentially making it easier to generate large numbers of cells for therapeutic means. In contrast, adult stem cell might not divide fast enough to offer immediate treatment.

- Embryonic stem cells have greater plasticity, potentially allowing them to treat a wider range of diseases.

- Adult stem cells from the patient's own body might not be effective in treatment of genetic disorders. Allogeneic embryonic stem cell transplantation (i.e. from a healthy donor) may be more practical in these cases than gene therapy of a patient's own cell.

- DNA abnormalities found in adult stem cells that are caused by toxins and sunlight may make them poorly suited for treatment.

- Embryonic stem cells have been shown to be effective in treating heart damage in mice.

- Embryonic stem cells have the potential to cure chronic and degenerative diseases which current medicine has been unable to effectively treat.

Individuality

- Before the primitive streak is formed when the embryo attaches to the uterus around 14 days after fertilization, two fertilized eggs can combine by fusing together and develop into one person (a tetragametic chimera). Since a fertilized egg has the potential to be two individuals or half of one, some believe it can only be considered a 'potential' person, not an actual one. Those who subscribe to this belief then hold that destroying a blastocyst for embryonic stem cells is ethical.

Viability

- Viability is another standard under which embryos and fetuses have been regarded as human lives. In the United States, the 1973 Supreme Court case of *Roe v. Wade* concluded that viability determined the permissibility of abortions performed for reasons other than the protection of the woman's health, defining viability as the point at which a fetus is "potentially able to live outside the mother's womb, albeit with artificial aid." The point of viability was 24 to 28 weeks when the case was decided and has since moved to about 22 weeks due to advancement in medical technology. Embryos used in medical research for stem cells are well below development that would enable viability.

Objection

Alternatives

This argument is used by opponents of embryonic destruction, as well as researchers specializing in adult stem cell research.

Pro-life supporters often claim that the use of adult stem cells from sources such as umbilical cord blood has consistently produced more promising results than the use of embryonic stem cells. Furthermore, adult stem cell research may be able to make greater advances if less money and resources were channeled into embryonic stem cell research.

In the past, it has been a necessity to research embryonic stem cells and in doing so destroy them for research to progress. As a result of the research done with both embryonic and adult stem cells, new techniques may make the necessity for embryonic cell research obsolete. Because many of the restrictions placed on stem cell research have been based on moral dilemmas surrounding the use of embryonic cells, there will likely be rapid advancement in the field as the techniques that created those issues are becoming less of a necessity. Many funding and research restrictions on embryonic cell research will not impact research on IPSCs (induced pluripotent stem cells) allowing for a promising portion of the field of research to continue relatively unhindered by the ethical issues of embryonic research.

Adult stem cells have provided many different therapies for illnesses such as Parkinson's disease, leukemia, multiple sclerosis, lupus, sickle-cell anemia, and heart damage, (to date, embryonic stem cells have also been used in treatment) Moreover, there have been many advances in adult stem cell research, including a recent study where pluripotent adult stem cells were manufactured from differentiated fibroblast by the addition of specific transcription factors. Newly created stem cells were developed into an embryo and were integrated into newborn mouse tissues, analogous to the properties of embryonic stem cells.

Government Policy Stances

Europe

Austria, Denmark, France, Germany, and Ireland do not allow the production of embryonic stem cell lines, but the creation of embryonic stem cell lines is permitted in Finland, Greece, the Netherlands, Sweden, and the United Kingdom.

United States

Origins

In 1973, *Roe v. Wade* legalized abortion in the United States. Five years later, the first successful human *in vitro* fertilization resulted in the birth of Louise Brown in England. These developments prompted the federal government to create regulations barring the use of federal funds for research that experimented on human embryos. In 1995, the NIH Human Embryo Research Panel advised the administration of President Bill Clinton to permit federal funding for research on embryos left over from *in vitro* fertility treatments and also recommended federal funding of research on embryos specifically created for experimentation. In response to the panel's recommendations, the Clinton administration, citing moral and ethical concerns, declined to fund research on embryos created solely for research purposes, but did agree to fund research on leftover embryos created by *in vitro* fertility treatments. At this point, the Congress intervened and passed the Dickey Amendment in 1995 (the final bill, which included the Dickey Amendment, was signed into law by Bill Clinton) which prohibited any federal funding for the Department of Health and Human Services be used for research that resulted in the destruction of an embryo regardless of the source of that embryo.

In 1998, privately funded research led to the breakthrough discovery of human embryonic stem cells (hESC). This prompted the Clinton administration to re-examine guidelines for federal funding of embryonic research. In 1999, the president's National Bioethics Advisory Commission recommended that hESC harvested from embryos discarded after *in vitro* fertility treatments, but not from embryos created expressly for experimentation, be eligible for federal funding. Though embryo destruction had been inevitable in the process of harvesting hESC in the past (this is no longer the case), the Clinton administration had decided that it would be permissible under the Dickey Amendment to fund hESC research as long as such research did not itself directly cause the destruction of an embryo. Therefore, HHS issued its proposed regulation concerning hESC funding in 2001. Enactment of the new guidelines was delayed by the incoming George W. Bush administration which decided to reconsider the issue.

President Bush announced, on August 9, 2001, that federal funds, for the first time, would be made available for hESC research on currently existing embryonic stem cell lines. President Bush authorized research on existing human embryonic stem cell lines, not on human embryos under a specific, unrealistic timeline in which the stem cell lines must have been developed. However, the Bush Administration chose not to permit taxpayer funding for research on hESC cell lines not currently in existence, thus limiting federal funding to research in which "the life-and-death decision has already been made". The Bush Administration's guidelines differ from the Clinton Administration guidelines which did not distinguish between currently existing and not-yet-existing hESC. Both the Bush and Clinton guidelines agree that the federal government should not fund hESC research that directly destroys embryos.

Neither Congress nor any administration has ever prohibited private funding of embryonic research. Public and private funding of research on adult and cord blood stem cells is unrestricted.

U.S. Congressional Response

In April 2004, 206 members of Congress signed a letter urging President Bush to expand federal funding of embryonic stem cell research beyond what Bush had already supported.

In May 2005, the House of Representatives voted 238-194 to loosen the limitations on federally funded embryonic stem-cell research — by allowing government-funded research on surplus frozen embryos from in vitro fertilization clinics to be used for stem cell research with the permission of donors — despite Bush's promise to veto the bill if passed. On July 29, 2005, Senate Majority Leader William H. Frist (R-TN), announced that he too favored loosening restrictions on federal funding of embryonic stem cell research. On July 18, 2006, the Senate passed three different bills concerning stem cell research. The Senate passed the first bill (Stem Cell Research Enhancement Act), 63-37, which would have made it legal for the federal government to spend federal money on embryonic stem cell research that uses embryos left over from *in vitro* fertilization procedures. On July 19, 2006 President Bush vetoed this bill. The second bill makes it illegal to create, grow, and abort fetuses for research purposes. The third bill would encourage research that would isolate pluripotent, i.e., embryonic-like, stem cells without the destruction of human embryos.

In 2005 and 2007, Congressman Ron Paul introduced the Cures Can Be Found Act, with 10 cosponsors. With an income tax credit, the bill favors research upon non embryonic stem cells obtained from placentas, umbilical cord blood, amniotic fluid, humans after birth, or unborn human offspring who died of natural causes; the bill was referred to committee. Paul argued that hESC research is outside of federal jurisdiction either to ban or to subsidize.

Bush vetoed another bill, the Stem Cell Research Enhancement Act of 2007, which would have amended the Public Health Service Act to provide for human embryonic stem cell research. The bill passed the Senate on April 11 by a vote of 63-34, then passed the House on June 7 by a vote of 247-176. President Bush vetoed the bill on July 19, 2007.

On March 9, 2009, President Obama removed the restriction on federal funding for newer stem cell lines. Two days after Obama removed the restriction, the president then signed the Omnibus Appropriations Act of 2009, which still contained the long-standing Dickey-Wicker provision which bans federal funding of "research in which a human embryo or embryos are destroyed, discarded, or knowingly subjected to risk of injury or death;" the Congressional provision effectively prevents federal funding being used to create new stem cell lines by many of the known methods. So, while scientists might not be free to create new lines with federal funding, President Obama's policy allows the potential of applying for such funding into research involving the hundreds of existing stem cell lines as well as any further lines created using private funds or state-level funding. The ability to apply for federal funding for stem cell lines created in the private sector is a significant expansion of options over the limits imposed by President Bush, who restricted funding to the 21 viable stem cell lines that were created before he announced his decision in 2001. The ethical concerns raised during Clinton's time in office continue to restrict hESC research and dozens of stem cell lines have been excluded from funding, now by judgment of an administrative office rather than presidential or legislative discretion.

Funding

In 2005, the NIH funded $607 million worth of stem cell research, of which $39 million was specifically used for hESC. Sigrid Fry-Revere has argued that private organizations, not the federal government, should provide funding for stem-cell research, so that shifts in public opinion and government policy would not bring valuable scientific research to a grinding halt.

In 2005, the State of California took out $3 billion in bond loans to fund embryonic stem cell research in that state.

Asia

China has one of the most permissive human embryonic stem cell policies in the world. In the absence of a public controversy, human embryo stem cell research is supported by policies that allow the use of human embryos and therapeutic cloning.

Religious Views

Jewish View

According to Rabbi Levi Yitzchak Halperin of the Institute for Science and Jewish Law in Jerusalem, embryonic stem cell research is permitted so long as it has not been implanted in the womb. Not only is it permitted, but research is encouraged, rather than wasting it.

> As long as it has not been implanted in the womb and it is still a frozen fertilized egg, it does not have the status of an embryo at all and there is no prohibition to destroy it... However in order to remove all doubt [as to the permissibility of destroying it], it is preferable not to destroy the pre-embryo unless it will otherwise not be implanted in the woman who gave the eggs (either because there are many fertilized eggs, or because one of the parties refuses to go on with the procedure - the husband or wife - or for any other reason). Certainly it should not be implanted into another woman.... The best and worthiest solution is to use it for life-saving purposes, such as for the treatment of people that suffered trauma to their nervous system, etc.
>
> — Rabbi Levi Yitzchak Halperin, Ma'aseh Choshev vol. 3, 2:6

Similarly, the sole Jewish majority state, Israel, permits research on embryonic stem cells.

Catholicism

The Catholic Church opposes human embryonic stem cell research calling it "an absolutely unacceptable act." The Church supports research that involves stem cells from adult tissues and the umbilical cord, as it "involves no harm to human beings at any state of development."

Baptists

The Southern Baptist Convention opposes human embryonic stem cell research on the grounds that "Bible teaches that human beings are made in the image and likeness of God (Gen. 1:27; 9:6) and protectable human life begins at fertilization." However, it supports adult stem cell research as it does "not require the destruction of embryos."

Methodism

The United Methodist Church opposes human embryonic stem cell research, saying, "a human embryo, even at its earliest stages, commands our reverence." However, it supports adult stem cell research, stating that there are "few moral questions" raised by this issue.

Pentecostalism

The Assemblies of God opposes human embryonic stem cell research, saying, it "perpetuates the evil of abortion and should be prohibited."

Islam

The religion of Islam favors the stance that scientific research and development in terms of stem cell research is allowed as long as it benefits society while using the least amount of harm to the subjects. "Stem cell research is one of the most controversial topics of our time period and has raised many religious and ethical questions regarding the research being done. With there being no true guidelines set forth in the Qur'an against the study of biomedical testing, Muslims have adopted any new studies as long as the studies do not contradict another teaching in the Qur'an. One of the teachings of the Qur'an states that "Whosoever saves the life of one, it shall be if he saves the life of humankind" (5:32), it is this teaching that makes stem cell research acceptable in the Muslim faith because of its promise of potential medical breakthrough."

The Church of Jesus Christ of Latter-day Saints

The First Presidency of The Church of Jesus Christ of Latter-day Saints "has not taken a position regarding the use of embryonic stem cells for research purposes. The absence of a position should not be interpreted as support for or opposition to any other statement made by Church members, whether they are for or against embryonic stem cell research."

Adult Stem Cell

Adult stem cells are undifferentiated cells, found throughout the body after development, that multiply by cell division to replenish dying cells and regenerate damaged tissues. Also known as somatic stem cells, they can be found in juvenile as well as adult animals and human bodies.

Scientific interest in adult stem cells is centered on their ability to divide or *self-renew* indefinitely, and generate all the cell types of the organ from which they originate, potentially regenerating the entire organ from a few cells. Unlike embryonic stem cells, the use of human adult stem cells in research and therapy is not considered to be controversial, as they are derived from adult tissue samples rather than human embryos designated for scientific research. They have mainly been studied in humans and model organisms such as mice and rats.

Defining Properties

A stem cell possesses two properties:

- Self-renewal, which is the ability to go through numerous cycles of cell division while still maintaining its undifferentiated state.

- *multipotency or multidifferentiative potential,* which is the ability to generate progeny of several distinct cell types, (for example glial cells and neurons) as opposed to unipotency, which is the term for cells that are restricted to producing a single-cell type. However, some researchers do not consider multipotency to be essential, and believe that unipotent self-renewing stem cells can exist. These properties can be illustrated with relative ease *in vitro,* using methods such as clonogenic assays, where the progeny of a single cell is characterized. However, it is known that *in vitro* cell culture conditions can alter the behavior of cells, proving that a particular subpopulation of cells possesses stem cell properties *in vivo* is challenging, and so considerable debate exists as to whether some proposed stem cell populations in the adult are indeed stem cells.

Lineage

To ensure the safety of others, stem cells undergo two types of cell division. Symmetric division gives rise to two identi-cal daughter cells, both endowed with stem cell properties, whereas asymmetric divi-sion produces only one of those stem cells and a progenitor cell with limited self-renew-al potential. Progenitors can go through several rounds of cell division before finally differentiating into a mature cell. It is believed that the molecular distinction between symmetric and asymmetric divisions lies in differential segregation of cell membrane proteins (such as receptors) between the daughter cells.

Multidrug Resistance

Adult stem cells express transporters of the ATP-binding cassette family that actively pump a diversity of organic molecules out of the cell. Many pharmaceuticals are exported by these transporters conferring multidrug resistance onto the cell. This complicates the design of drugs, for instance neural stem cell targeted therapies for the treatment of clinical depression.

Signaling Pathways

Adult stem cell research has been focused on uncovering the general molecular mechanisms that control their self-renewal and differentiation.

- Notch

The Notch pathway has been known to developmental biologists for decades. Its role in control of stem cell proliferation has now been demonstrated for several cell types including haematopoietic, neural, and mammary stem cells.

- Wnt

These developmental pathways are also strongly implicated as stem cell regulators.

- TGFβ

The TGFβ family of cytokines regulate the stemness of both normal and cancer stem cells.

Plasticity

Discoveries in recent years have suggested that adult stem cells might have the ability to differentiate into cell types from different germ layers. For instance, neural stem cells from the brain, which are derived from ectoderm, can differentiate into ectoderm, mesoderm, and endoderm. Stem cells from the bone marrow, which is derived from mesoderm, can differentiate into liver, lung, GI tract and skin, which are derived from endoderm and mesoderm. This phenomenon is referred to as stem cell transdifferentiation or plasticity. It can be induced by modifying the growth medium when stem cells are cultured *in vitro* or transplanting them to an organ of the body different from the one they were originally isolated from. There is yet no consensus among biologists on the prevalence and physiological and therapeutic relevance of stem cell plasticity. More recent findings suggest that pluripotent stem cells may reside in blood and adult tissues in a dormant state. These cells are referred to as "Blastomere Like Stem Cells" (Am Surg. 2007 Nov;73:1106-10) and "very small embryonic like" - "VSEL" stem cells, and display pluripotency in vitro. As BLSC's and VSEL cells are present in virtually all adult tissues, including lung, brain, kidneys, muscles, and pancreas Co-purification of BLSC's and VSEL cells with other populations of adult stem cells may explain the apparent pluripotency of adult stem cell populations. However, recent studies have shown that both human and murine VSEL cells lack stem cell characteristics and are not pluripotent.

Aging

Stem cell function becomes impaired with age, and this contributes to progressive deterioration of tissue maintenance and repair. A likely important cause of increasing stem cell dysfunction is age-dependent accumulation of DNA damage in both stem cells and the cells that comprise the stem cell environment.

Types

Hematopoietic Stem Cells

Hematopoietic stem cells are found in the bone marrow and umbilical cord blood and give rise to all the blood cell types.

Mammary Stem Cells

Mammary stem cells provide the source of cells for growth of the mammary gland during puberty and gestation and play an important role in carcinogenesis of the breast. Mammary stem cells have been isolated from human and mouse tissue as well as from cell lines derived from the mammary gland. Single such cells can give rise to both the luminal and myoepithelial cell types of the gland, and have been shown to have the ability to regenerate the entire organ in mice.

Intestinal Stem Cells

Intestinal stem cells divide continuously throughout life and use a complex genetic program to produce the cells lining the surface of the small and large intestines. Intestinal stem cells reside near the base of the stem cell niche, called the crypts of Lieberkuhn. Intestinal stem cells are probably the source of most cancers of the small intestine and colon.

Mesenchymal Stem Cells

Mesenchymal stem cells (MSCs) are of stromal origin and may differentiate into a variety of tissues. MSCs have been isolated from placenta, adipose tissue, lung, bone marrow and blood, Wharton's jelly from the umbilical cord, and teeth (perivascular niche of dental pulp and periodontal ligament). MSCs are attractive for clinical therapy due to their ability to differentiate, provide trophic support, and modulate innate immune response.

Endothelial Stem Cells

Endothelial stem cells are one of the three types of multipotent stem cells found in the bone marrow. They are a rare and controversial group with the ability to differentiate into endothelial cells, the cells that line blood vessels.

Neural Stem Cells

The existence of stem cells in the adult brain has been postulated following the discovery that the process of neurogenesis, the birth of new neurons, continues into adulthood in rats. The presence of stem cells in the mature primate brain was first reported in 1967. It has since been shown that new neurons are generated in adult mice, songbirds and primates, including humans. Normally, adult neurogenesis is restricted to two areas of the brain – the subventricular zone, which lines the lateral ventricles, and the dentate gyrus of the hippocampal formation. Although the generation of new neurons in the hippocampus is well established, the presence of true self-renewing stem cells there has been debated. Under certain circumstances, such as following tissue damage in ischemia, neurogenesis can be induced in other brain regions, including the neocortex.

Neural stem cells are commonly cultured *in vitro* as so called neurospheres – floating heterogeneous aggregates of cells, containing a large proportion of stem cells. They can be propagated for extended periods of time and differentiated into both neuronal and glia cells, and therefore behave as stem cells. However, some recent studies suggest that this behaviour is induced by the culture conditions in progenitor cells, the progeny of stem cell division that normally undergo a strictly limited number of replication cycles *in vivo*. Furthermore, neurosphere-derived cells do not behave as stem cells when transplanted back into the brain.

Neural stem cells share many properties with haematopoietic stem cells (HSCs). Remarkably, when injected into the blood, neurosphere-derived cells differentiate into various cell types of the immune system.

Olfactory Adult Stem Cells

Olfactory adult stem cells have been successfully harvested from the human olfactory mucosa cells, which are found in the lining of the nose and are involved in the sense of smell. If they are given the right chemical environment these cells have the same ability as embryonic stem cells to develop into many different cell types. Olfactory stem cells hold the potential for therapeutic applications and, in contrast to neural stem cells, can be harvested with ease without harm to the patient. This means they can be easily obtained from all individuals, including older patients who might be most in need of stem cell therapies.

Neural Crest Stem Cells

Hair follicles contain two types of stem cells, one of which appears to represent a remnant of the stem cells of the embryonic neural crest. Similar cells have been found in the gastrointestinal tract, sciatic nerve, cardiac outflow tract and spinal and sympathetic ganglia. These cells can generate neurons, Schwann cells, myofibroblast, chondrocytes and melanocytes.

Testicular Cells

Multipotent stem cells with a claimed equivalency to embryonic stem cells have been derived from spermatogonial progenitor cells found in the testicles of laboratory mice by scientists in Germany and the United States, and, a year later, researchers from Germany and the United Kingdom confirmed the same capability using cells from the testicles of humans. The extracted stem cells are known as human adult germline stem cells (GSCs)

Multipotent stem cells have also been derived from germ cells found in human testicles.

Adult Stem Cell Therapies

The therapeutic potential of adult stem cells is the focus of much scientific research, due to their ability to be harvested from the patient. In common with embryonic stem cells, adult stem cells have the ability to differentiate into more than one cell type, but unlike the former they are often restricted to certain types or "lineages". The ability of a differentiated stem cell of one lineage to produce cells of a different lineage is called transdifferentiation. Some types of adult stem cells are more capable of transdifferentiation than others, but for many there is no evidence that such a transformation is possible. Consequently, adult stem therapies require a stem cell source of the specific lineage needed, and harvesting and/or culturing them up to the numbers required is a challenge. Additionally, cues from the immediate environment (including how stiff or porous the surrounding structure/extracellular matrix is) can alter or enhance the fate and differentiation of the stem cells.

Sources

Pluripotent stem cells, i.e. cells that can give rise to any fetal or adult cell type, can be found in a number of tissues, including umbilical cord blood. Using genetic reprogramming, pluripotent stem cells equivalent to embryonic stem cells have been derived from human adult skin tissue. Other adult stem cells are multipotent, meaning they are restricted in the types of cell they can become, and are generally referred to by their tissue origin (such as mesenchymal stem cell, adipose-derived stem cell, endothelial stem cell, etc.). A great deal of adult stem cell research has focused on investigating their capacity to divide or self-renew indefinitely, and their potential for differentiation. In mice, pluripotent stem cells can be directly generated from adult fibroblast cultures.

Clinical Applications

Adult stem cell treatments have been used for many years to successfully treat leukemia and related bone/blood cancers utilizing bone marrow transplants. The use of adult stem cells in research and therapy is not considered as controversial as the use of embryonic stem cells, because the production of adult stem cells does not require the destruction of an embryo.

Early regenerative applications of adult stem cells has focused on intravenous delivery of blood progenitors known as Hematopetic Stem Cells (HSC's). CD34+ hematopoietic Stem Cells have been clinically applied to treat various diseases including spinal cord injury, liver cirrhosis and Peripheral Vascular disease. Research has shown that CD34+ hematopoietic Stem Cells are relatively more numerous in men than in women of reproductive age group among spinal cord Injury victims. Other early commercial applications have focused on Mesenchymal Stem Cells (MSCs). For both cell lines, direct injection or placement of cells into a site in need of repair may be the preferred method of treatment, as vascular delivery suffers from a "pulmonary first pass effect" where intravenous injected cells are sequestered in the lungs. Clinical case reports in orthopedic applications have been published. Wakitani has published a small case series of nine defects in five knees involving surgical transplantation of mesenchymal stem cells with coverage of the treated chondral defects. Centeno et al. have reported high field

MRI evidence of increased cartilage and meniscus volume in individual human clinical subjects as well as a large n=227 safety study. Many other stem cell based treatments are operating outside the US, with much controversy being reported regarding these treatments as some feel more regulation is needed as clinics tend to exaggerate claims of success and minimize or omit risks.

Adult Stem Cells and Cancer

In recent years, acceptance of the concept of adult stem cells has increased. There is now a hypothesis that stem cells reside in many adult tissues and that these unique reservoirs of cells not only are responsible for the normal reparative and regenerative processes but are also considered to be a prime target for genetic and epigenetic chang-es, culminating in many abnormal conditions including cancer.

Cancer Stem Cell

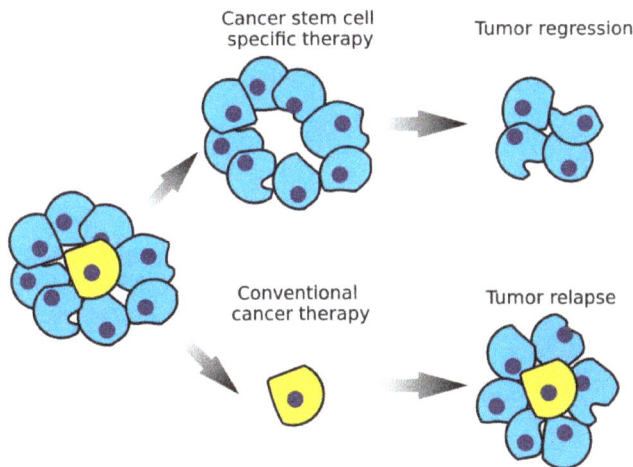

Figure 1: Stem cell specific and conventional cancer therapies

Cancer stem cells (CSCs) are cancer cells (found within tumors or hematological cancers) that possess characteristics associated with normal stem cells, specifically the ability to give rise to all cell types found in a particular cancer sample. CSCs are therefore tumorigenic (tumor-forming), perhaps in contrast to other non-tumorigenic cancer cells. CSCs may generate tumors through the stem cell processes of self-renewal and differentiation into multiple cell types. Such cells are hypothesized to persist in tumors as a distinct population and cause relapse and metastasis by giving rise to new tumors. Therefore, development of specific therapies targeted at CSCs holds hope for improvement of survival and quality of life of cancer patients, especially for patients with metastatic disease.

Existing cancer treatments have mostly been developed based on animal models, where therapies able to promote tumor shrinkage were deemed effective. However, animals do not provide a complete model of human disease. In particular, in mice, whose life spans do not exceed two years, tumor relapse is difficult to study.

The efficacy of cancer treatments is, in the initial stages of testing, often measured by the ablation fraction of tumor mass (fractional kill). As CSCs form a small proportion of the tumor, this may not necessarily select for drugs that act specifically on the stem cells. The theory suggests that conventional chemotherapies kill differentiated or differentiating cells, which form the bulk of the tumor but do not generate new cells. A population of CSCs, which gave rise to it, could remain untouched and cause relapse.

Cancer stem cells were first identified by John Dick in acute myeloid leukemia in the late 1990s. Since the early 2000s they have been an intense cancer research focus.

Tumor Propagation Models

In different tumor subtypes, cells within the tumor population exhibit functional heterogeneity and tumors are formed from cells with various proliferative and differentiation capacities. This functional heterogeneity among cancer cells has led to the creation of multiple propagation models to account for heterogeneity and differences in tumor-regenerative capacity: the cancer stem cell (CSC) and clonal evolution models

The cancer stem cell model refers to a subset of tumor cells that have the ability to self-renew and are able to generate the diverse tumor cells. These cells have been termed cancer stem cells to reflect their stem-like properties. One implication of the CSC model and the existence of CSCs is that the tumor population is hierarchically arranged with CSCs lying at the apex (Fig. 3).

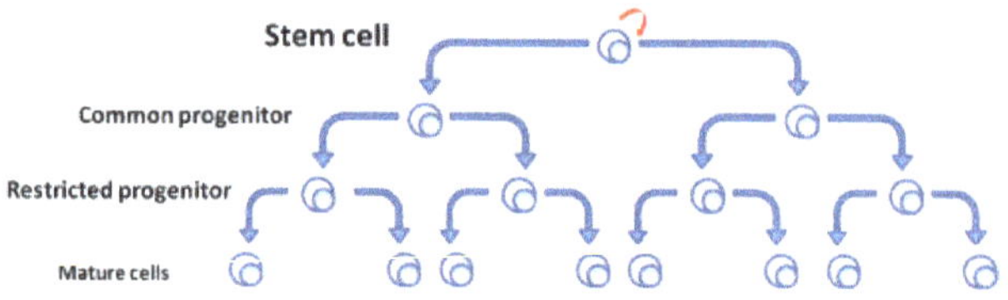

Figure 2: A normal cellular hierarchy comprising stem cells at the apex, which generate common and more restricted progenitor cells and ultimately the mature cell types that constitute particular tissues.

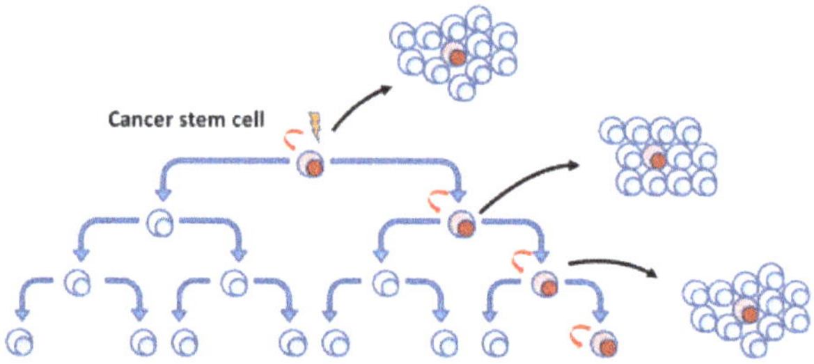

Figure 3. In the cancer stem cell (CSC) model, only the CSCs have the ability to generate a tumor, based on their self-renewal properties and proliferative potential.

The clonal evolution model postulates that mutant tumor cells with a growth advantage outproliferate others. Cells in the dominant population have a similar potential for initiating tumor growth (Fig. 4).

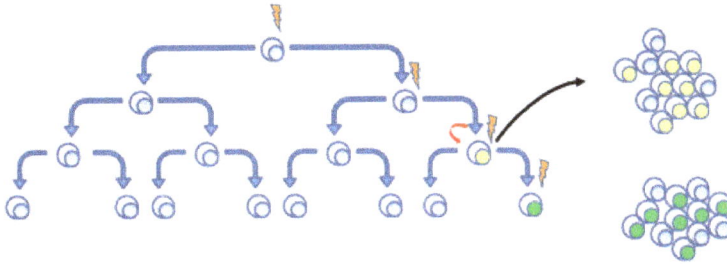

Figure 4: In the clonal evolution model, all undifferentiated cells have similar possibility to change into a tumorigenic cell.

These two models are not mutually exclusive, as CSCs themselves undergo clonal evolution. Thus, the secondary more dominant CSCs may emerge, if a mutation confers more aggressive properties (Fig. 5).

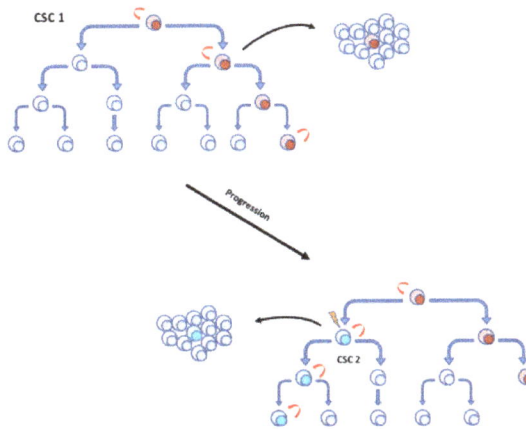

Figure 5: Both tumor models may play a role in the maintenance of a tumor. Initially, tumor growth is assured with a specific CSC (CSC1). With tumor progression, another CSC (CSC 2) may arise due the clonal selection. The development of a new more aggressive CSC may result from the acquisition of an additional mutation or epigenetic modification.

Debate

The existence of CSCs is under debate, because many studies found no cells with their specific characteristics. Cancer cells must be capable of continuous proliferation and self-renewal to retain the many mutations required for carcinogenesis and to sustain the growth of a tumor, since differentiated cells (constrained by the Hayflick Limit) cannot divide indefinitely. If most tumor cells are endowed with stem cell properties, targeting tumor size directly is a valid strategy. If they are a small minority, targeting them may be more effective. Another debate is over the origin of CSCs - whether from disregulation of normal stem cells or from a more specialized population that acquired the ability to self-renew (which is related to the issue of stem cell plasticity).

Evidence

The first conclusive evidence for CSCs came in 1997. Bonnet and Dick isolated a subpopulation of leukemia cells that expressed surface marker CD34, but not CD38. The authors established that the CD34$^+$/CD38$^-$ subpopulation is capable of initiating tumors in NOD/SCID mice that were histologically similar to the donor. The first evidence of a solid tumor cancer stem-like cell followed in 2002 with the discovery of a clonogenic, sphere-forming cell isolated and characterized from human brain gliomas. Human cortical glial tumors contain neural stem-like cells expressing astroglial and neuronal markers *in vitro*.

In cancer research experiments, tumor cells are sometimes injected into an experimental animal to establish a tumor. Disease progression is then followed in time and novel drugs can be tested for their efficacy. Tumor formation requires thousands or tens of thousands of cells to be introduced. Classically, this was explained by poor methodology (i.e., the tumor cells lose their viability during transfer) or the critical importance of the microenvironment, the particular biochemical surroundings of the injected cells. Supporters of the CSC paradigm argue that only a small fraction of the injected cells, the CSCs, have the potential to generate a tumor. In human acute myeloid leukemia the frequency of these cells is less than 1 in 10,000.

Further evidence comes from histology. Many tumors are heterogeneous and contain multiple cell types native to the host organ. Heterogeneity is commonly retained by tumor metastases. This suggests that the cell that produced them had the capacity to generate multiple cell types, a classical hallmark of stem cells.

The existence of leukemia stem cells prompted research into other cancers. CSCs have recently been identified in several solid tumors, including:

- Brain
- Breast
- Colon
- Ovary
- Pancreas
- Prostate
- Melanoma
- Multiple Myeloma
- Non-melanoma skin cancer

Mechanistic and Mathematical Models

Once the pathways to cancer are hypothesized, it is possible to develop predictive mathematical models, e.g., based on the cell compartment method. For instance, the growths of abnormal cells can be denoted with specific mutation probabilities. Such a model predicted that repeated insult to mature cells increases the formation of abnormal progeny and the risk of cancer. The clinical efficacy of such models remains unestablished.

Origin

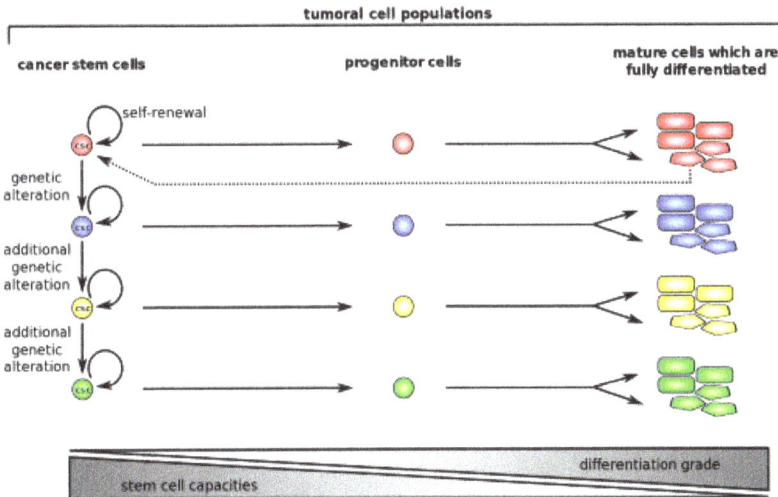

Figure 6: Hierarchical organisation of a tumour according to the CSC model

The origin of CSCs is an active research area. The answer may depend on the tumor type and phenotype. So far the hypothesis that tumors originate from a single "cell of origin" has not been demonstrated using the cancer stem cell model. This is because cancer stem cells are not present in end-stage tumors.

Origin hypotheses include mutants in developing stem or progenitor cells, mutants in adult stem cells or adult progenitor cells and mutant, differentiated cells that acquire stem-like attributes. These theories often focus on a tumor's "cell of origin".

Hypotheses

Stem Cell Mutation

The "mutation in stem cell niche populations during development" hypothesis claims that these developing stem populations are mutated and then reproduce so that the mutation is shared by many descendants. These daughter cells are much closer to becoming tumors and their numbers increase the chance of a cancerous mutation.

Adult Stem Cells

Another theory associates adult stem cells (ASC) with tumor formation. This is most often associated with tissues with a high rate of cell turnover (such as the skin or gut). In these tissues, ASCs are candidates because of their frequent cell divisions (compared to most ASCs) in conjunction with the long lifespan of ASCs. This combination creates the ideal set of circumstances for mutations to accumulate: mutation accumulation is the primary factor that drives cancer initiation. Evidence shows that the association represents an actual phenomenon, although specific cancers have been linked to a specific cause.

De-differentiation

De-differentiation of mutated cells may create stem cell-like characteristics, suggesting that any cell might become a cancer stem cell. In other words, a fully differentiated cell undergoes several mutations that drive it back to a stem-like state.

Hierarchy

The concept of tumor hierarchy claims that a tumor is a heterogeneous population of mutant cells, all of which share some mutations, but vary in specific phenotype. A tumor hosts several types of stem cells, one optimal to the specific environment and other less successful lines. These secondary lines may be more successful in other environments, allowing the tumor to adapt, including adaptation to therapeutic intervention. If correct, this concept impacts cancer stem cell-specific treatment regimes. Such a hierarchy would complicate attempts to pinpoint the origin.

Identification

CSCs, now reported in most human tumors, are commonly identified and enriched using strategies for identifying normal stem cells that are similar across studies. These procedures include fluorescence-activated cell sorting (FACS), with antibodies directed at cell-surface markers and functional approaches including side population assay or Aldefluor assay. The CSC-enriched result is then implanted, at various doses, in immune-deficient mice to assess its tumor development capacity. This *in vivo* assay is called a limiting dilution assay. The tumor cell subsets that can initiate tumor development at low cell numbers are further tested for self-renewal capacity in serial tumor studies.

CSC can also be identified by efflux of incorporated Hoechst dyes via multidrug resistance (MDR) and ATP-binding cassette (ABC) Transporters.

Another approach is sphere-forming assays. Many normal stem cells such as hematopoietics or stem cells from tissues, under special culture conditions, form three-dimensional spheres that can differentiate. As with normal stem cells, the CSCs isolated from brain or prostate tumors also have the ability to form anchor-independent spheres.

Heterogeneity (Markers)

CSCs have been identified in various solid tumors. Markers specific for normal stem cells are commonly used for isolating CSCs from solid and hematological tumors. Cell surface markers have proved useful for isolation of CSC-enriched populations including CD133 (also known as PROM1), CD44, CD24, EpCAM (epithelial cell adhesion molecule, also known as epithelial specific antigen, ESA), THY1, ATP-binding cassette B5 (ABCB5)., and CD200.

CD133 (prominin 1) is a five-transmembrane domain glycoprotein expressed on $CD34^+$ stem and progenitor cells, in endothelial precursors and fetal neural stem cells. It has been detected using its glycosylated epitope known as AC133.

EpCAM (epithelial cell adhesion molecule, ESA, TROP1) is hemophilic Ca^{2+}-independent cell adhesion molecule expressed on the basolateral surface of most epithelial cells.

CD90 (THY1) is a glycosylphosphatidylinositol glycoprotein anchored in the plasma membrane and involved in signal transduction. It may also mediate adhesion between thymocytes and thymic stroma.

CD44 (PGP1) is an adhesion molecule that has pleiotropic roles in cell signaling, migration and homing. It has multiple isoforms, including CD44H, which exhibits high affinity for hyaluronate and CD44V which has metastatic properties.

CD24 (HSA) is a glycosylated glycosylphosphatidylinositol-anchored adhesion molecule, which has co-stimulatory role in B and T cells.

CD200 (OX-2) is a type 1 membrane glycoprotein, which delivers an inhibitory signal to immune cells including T cells, natural killer cells and macrophages.

ALDH is a ubiquitous aldehyde dehydrogenase family of enzymes, which catalyzes the oxidation of aromatic aldehydes to carboxyl acids. For instance, it has a role in conversion of retinol to retinoic acid, which is essential for survival.

The first solid malignancy from which CSCs were isolated and identified was breast cancer and they are the most intensely studied. Breast CSCs have been enriched in $CD44^+CD24^{-/low}$, SP and $ALDH^+$ subpopulations. Breast CSCs are apparently phenotypically diverse. CSC marker expression in breast cancer cells is apparently heterogeneous and breast CSC populations vary across tumors. Both $CD44^+CD24^-$ and $CD44^+CD24^+$ cell populations are tumor initiating cells; however, CSC are most highly enriched using the marker profile $CD44^+CD49f^{hi}CD133/2^{hi}$.

CSCs have been reported in many brain tumors. Stem-like tumor cells have been identified using cell surface markers including CD133, SSEA-1 (stage-specific embryonic antigen-1), EGFR and CD44. The use of CD133 for identification of brain tumor stem-like cells may be problematic because tumorigenic cells are found in both $CD133^+$ and $CD133^-$ cells in some gliomas and some $CD133^+$ brain tumor cells may not possess tumor-initiating capacity.

CSCs were reported in human colon cancer. For their identification, cell surface markers such as CD133, CD44 and ABCB5, functional analysis including clonal analysis and

Aldefluor assay were used. Using CD133 as a positive marker for colon CSCs generated conflicting results. The AC133 epitope, but not the CD133 protein, is specifically expressed in colon CSCs and its expression is lost upon differentiation. In addition, CD44$^+$ colon cancer cells and additional sub-fractionation of CD44$^+$EpCAM$^+$ cell population with CD166 enhance the success of tumor engraftments.

Multiple CSCs have been reported in prostate, lung and many other organs, including liver, pancreas, kidney or ovary. In prostate cancer, the tumor-initiating cells have been identified in CD44$^+$ cell subset as CD44$^+\alpha2\beta1^+$, TRA-1-60$^+$CD151$^+$CD166$^+$ or ALDH$^+$ cell populations. Putative markers for lung CSCs have been reported, including CD133$^+$, ALDH$^+$, CD44$^+$ and oncofetal protein 5T4$^+$.

Metastasis

Metastasis is the major cause of tumor lethality. However, not every tumor cell can metastasize. This potential depends on factors that determine growth, angiogenesis, invasion and other basic processes.

Epithelial-mesenchymal Transition

In epithelial tumors, the epithelial-mesenchymal transition (EMT) is considered to be a crucial event. EMT and the reverse transition from mesenchymal to an epithelial phenotype (MET) are involved in embryonic development, which involves disruption of epithelial cell homeostasis and the acquisition of a migratory mesenchymal phenotype. EMT appears to be controlled by canonical pathways such as WNT and transforming growth factor β.

EMT's important feature is the loss of membrane E-cadherin in adherens junctions, where β-catenin may play a significant role. Translocation of β-catenin from adherens junctions to the nucleus may lead to a loss of E-cadherin and subsequently to EMT. Nuclear β-catenin apparently can directly, transcriptionally activate EMT-associated target genes, such as the E-cadherin gene repressor SLUG (also known as SNAI2). Mechanical properties of the tumor microenvironment, or hypoxia, can influence CSC properties and metastatic behavior.

Tumor cells undergoing an EMT may be precursors for metastatic cancer cells, or even metastatic CSCs. In the invasive edge of pancreatic carcinoma, a subset of CD133$^+$CXCR4$^+$ (receptor for CXCL12 chemokine also known as a SDF1 ligand) cells was defined. These cells exhibited significantly stronger migratory activity than their counterpart CD133$^+$CXCR4$^-$ cells, but both showed similar tumor development capacity. Moreover, inhibition of the CXCR4 receptor reduced metastatic potential without altering tumorigenic capacity.

Two-phase Expression Pattern

In breast cancer CD44$^+$CD24$^{-/low}$ cells are detectable in metastatic pleural effusions. By contrast, an increased number of CD24$^+$ cells have been identified in distant metastases in breast cancer patients. It is possible that CD44$^+$CD24$^{-/low}$ cells initially metastasize and in the new site change their phenotype and undergo limited differentiation. The two-phase expression pattern hypothesis proposes two forms of cancer stem cells - stationary (SCS) and mobile (MCS). SCS are embedded in tissue and persist in differentiated areas throughout tumor progression. MCS are located at the tumor-host interface. These cells are apparently derived from SCS through the acquisition of transient EMT (Figure 7).

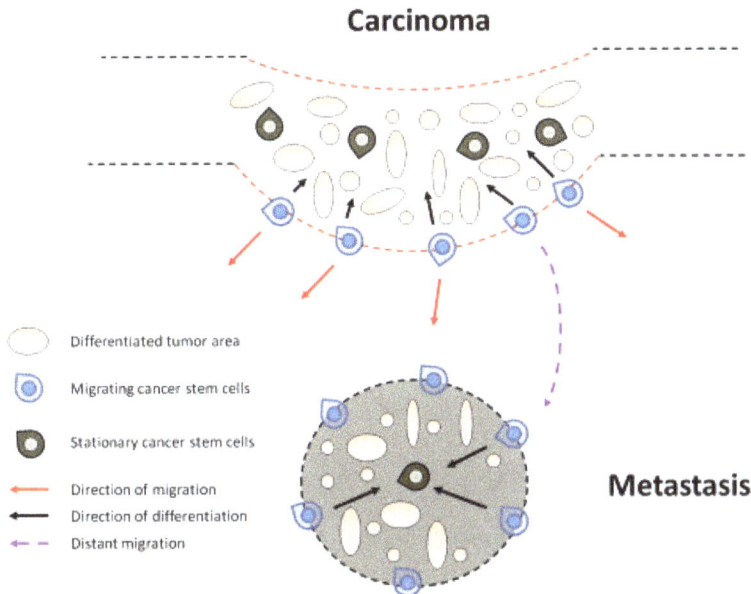

Figure 7: The concept of migrating cancer stem cells (MSC). Stationary cancer stem cells are embedded in begin carcinomas and these cells are detectable in the differentiated central area of a tumor. The important step toward malignancy is the induction of epithelial mesenchymal transition (EMT) in the stationary cancer stem cells (SCS), which become mobile or migrating cancer stem cells. MCS cells divide asymmetrically. One daughter cell starts proliferation and differentiation. The remaining MCS migrates a short distance before undergoing new asymmetric division, or starts dissemination through blood vessela or lymphatic vessels and produces a metastasis.

Implications

CSCs have implications for cancer therapy, including for disease identification, selective drug targets, prevention of metastasis and intervention strategies.

Treatment

Somatic stem cells are naturally resistant to chemotherapeutic agents. They produce various pumps (such as MDR) that pump out drugs and DNA repair proteins. They have a slow rate of cell turnover (chemotherapeutic agents naturally target rapidly replicating cells). CSCs that develop from normal stem cells may also produce these proteins, which could increase their resistance towards chemotherapy. The surviving CSCs then repopulate the tumor, causing a relapse.

Targeting

Selectively targeting CSCs may allow treatment of aggressive, non-resectable tumors, as well as prevent metastasis and relapse. The hypothesis suggests that upon CSC elimination, cancer could regress due to differentiation and/or cell death. The fraction of tumor cells that are CSCs and therefore need to be eliminated is unclear.

Studies looked for specific markers and for proteomic and genomic tumor signatures that distinguish CSCs from others. In 2009, scientists identified the compound salinomycin, which selectively reduces the proportion of breast CSCs in mice by more than 100-fold relative to Paclitaxel, a commonly used chemotherapeutic agent. Some types of cancer cells can survive treatment with salinomycin through autophagy, whereby cells use acidic organelles such as lysosomes to degrade and recycle certain types of proteins. The use of autophagy inhibitors can kill cancer stem cells that survive by autophagy.

The cell surface receptor interleukin-3 receptor-alpha (CD123) is overexpressed on CD34+CD38- leukemic stem cells (LSCs) in acute myelogenous leukemia (AML) but not on normal CD34+CD38- bone marrow cells. Treating AML-engrafted NOD/SCID mice with a CD123-specific monoclonal antibody impaired LSCs homing to the bone marrow and reduced overall AML cell repopulation including the proportion of LSCs in secondary mouse recipients.

A 2015 study packaged nanoparticles with miR-34a and ammonium bicarbonate and delivered them to prostate CSCs in a mouse model. Then they irradiated the area with near-infrared laser light. This caused the nanoparticles to swell three times or more in size bursting the endosomes and dispersing the RNA in the cell. miR-34a can lower the levels of CD44.

Pathways

The design of new drugs for targeting CSCs requires understanding the cellular mechanisms that regulate cell proliferation. The first advances in this area were made with hematopoietic stem cells (HSCs) and their transformed counterparts in leukemia, the disease for which the origin of CSCs is best understood. Stem cells of many organs share the same cellular pathways as leukemia-derived HSCs.

A normal stem cell may be transformed into a CSC through disregulation of the proliferation and differentiation pathways controlling it or by inducing oncoprotein activity.

BMI-1

The Polycomb group transcriptional repressor Bmi-1 was discovered as a common oncogene activated in lymphoma and later shown to regulate HSCs. The role of Bmi-1 has been illustrated in neural stem cells. The pathway appears to be active in CSCs of pediatric brain tumors.

Notch

The Notch pathway plays a role in controlling stem cell proliferation for several cell types including hematopoietic, neural and mammary SCs. Components of this pathway have been proposed to act as oncogenes in mammary and other tumors.

A branch of the Notch signaling pathway that involves the transcription factor Hes3 regulates a number of cultured cells with CSC characteristics obtained from glioblastoma patients.

Sonic Hedgehog and Wnt

These developmental pathways are SC regulators. Both Sonic hedgehog (SHH) and Wnt pathways are commonly hyperactivated in tumors and are necessary to sustain tumor growth. However, the Gli transcription factors that are regulated by SHH take their name from gliomas, where they are highly expressed. A degree of crosstalk exists between the two pathways and they are commonly activated together. By contrast, in colon cancer hedgehog signalling appears to antagonise Wnt.

Sonic hedgehog blockers are available, such as cyclopamine. A water-soluble cyclopamine may be more effective in cancer treatment. DMAPT, a water-soluble derivative of parthenolide, induces oxidative stress and inhibits NF-κB signaling for AML (leukemia) and possibly myeloma and prostate cancer. Telomerase is a study subject in CSC physiology. GRN163L (Imetelstat) was recently started in trials to target myeloma stem cells.

Wnt signaling can become independent of regular stimuli, through mutations in downstream oncogenes and tumor suppressor genes that become permanently activated even though the normal receptor has not received a signal. β-catenin binds to transcription factors such as the protein TCF4 and in combination the molecules activate the necessary genes. LF3 strongly inhibits this binding *in vitro,* in cell lines and reduced tumor growth in mouse models. It prevented replication and reduced their ability to migrate, all without affecting healthy cells. No cancer stem cells remained after treatment. The discovery was the product of "rational drug design", involving AlphaScreens and ELISA technologies.

Cancer Stem Cells Spheroids (3D Module)

The monolayer of CSCs grown as spheroids showed better growth rate than MDA-MB 231 cells, which shows the efficacy of the 3D spheroid format of CSCs. CD44 shows increased expression in spheroids compared to 2D culture of MDA-MB 231. ALDH1 iskey marker of breast stem cells wa.sIt highly expressed in BCSCs and MDA-MB 231 grown in 3D, while they are absent in CSCs and MDA-MB 231 cells grown in 2D.

Induced Pluripotent Stem Cell

Induced pluripotent stem cells (also known as iPS cells or iPSCs) are a type of pluripotent stem cell that can be generated directly from adult cells. The iPSC technology was pioneered by Shinya Yamanaka's lab in Kyoto, Japan, who showed in 2006 that the introduction of four specific genes encoding transcription factors could convert adult cells into pluripotent stem cells. He was awarded the 2012 Nobel Prize along with Sir John Gurdon "for the discovery that mature cells can be reprogrammed to become pluripotent."

Pluripotent stem cells hold great promise in the field of regenerative medicine. Because they can propagate indefinitely, as well as give rise to every other cell type in the body (such as neurons, heart, pancreatic, and liver cells), they represent a single source of cells that could be used to replace those lost to damage or disease.

The most well-known type of pluripotent stem cell is the embryonic stem cell. However, since the generation of embryonic stem cells involves destruction (or at least manipulation) of the pre-implantation stage embryo, there has been much controversy surrounding their use. Further, because embryonic stem cells can only be derived from embryos, it has so far not been feasible to create patient-matched embryonic stem cell lines.

Since iPSCs can be derived directly from adult tissues, they not only bypass the need for embryos, but can be made in a patient-matched manner, which means that each individual could have their own pluripotent stem cell line. These unlimited supplies of autologous cells could be used to generate transplants without the risk of immune rejection. While the iPSC technology has not yet advanced to a stage where therapeutic transplants have been deemed safe, iPSCs are readily being used in personalized drug discovery efforts and understanding the patient-specific basis of disease.Depending on the methods used, reprogramming of adult cells to obtain iPSCs may pose significant risks that could limit their use in humans. For example, if viruses are used to genomically alter the cells, the expression of oncogenes (cancer-causing genes) may potentially be triggered. In February 2008, scientists announced the discovery of a technique that could remove oncogenes after the induction of pluripotency, thereby increasing the potential use of iPS cells in human diseases.

Production

iPSCs are typically derived by introducing products of specific set of pluripotency-associated genes, or "reprogramming factors", into a given cell type. The original set of reprogramming factors (also dubbed Yamanaka factors) are the transcription factors Oct4 (Pou5f1), Sox2, cMyc, and Klf4. While this combination is most conventional in producing iPSCs, each of the factors can be functionally replaced by related transcription factors, miRNAs, small molecules, or even non-related genes such as lineage specifiers.

A scheme of the generation of induced pluripotent stem (IPS) cells. (1)Isolate and culture donor cells. (2) Transduce stem cell-associated genes into the cells by viral vectors. Red cells indicate the cells expressing the exogenous genes. (3)Harvest and culture the cells according to ES cell culture, using mitotically inactivated feeder cells (lightgray). (4)A small subset of the transfected cells become iPS cells and generate ES-like colonies.

iPSC derivation is typically a slow and inefficient process, taking 1–2 weeks for mouse cells and 3–4 weeks for human cells, with efficiencies around 0.01%–0.1%. However, considerable advances have been made in improving the efficiency and the time it takes to obtain iPSCs. Upon introduction of reprogramming factors, cells begin to form colonies that resemble pluripotent stem cells, which can be isolated based on their morphology, conditions that select for their growth, or through expression of surface markers or reporter genes.

First Generation (Mouse)

Induced pluripotent stem cells were first generated by Shinya Yamanaka's team at Kyoto University, Japan, in 2006. They hypothesized that genes important to embryonic stem cell (ESC) function might be able to induce an embryonic state in adult cells. They chose twenty-four genes previously identified as important in ESCs and used retroviruses to deliver these genes to mouse fibroblasts. The fibroblasts were engineered so that any cells reactivating the ESC-specific gene, Fbx15, could be isolated using antibiotic selection.

Upon delivery of all twenty-four factors, ESC-like colonies emerged that reactivated the Fbx15 reporter and could propagate indefinitely. To identify the genes necessary for reprogramming, the researchers removed one factor at a time from the pool of twenty-four. By this process, they identified four factors, Oct4, Sox2, cMyc, and Klf4, which were each necessary and together sufficient to generate ESC-like colonies under selection for reactivation of Fbx15.

Similar to ESCs, these iPSCs had unlimited self-renewal and were pluripotent, contributing to lineages from all three germ layers in the context of embryoid bodies, teratomas, and fetal chimeras. However, the molecular makeup of these cells, including gene expression and epigenetic marks, was somewhere between that of a fibroblast and an ESC, and the cells failed to produce viable chimeras when injected into developing embryos.

Second Generation (Mouse)

In June 2007, three separate research groups, including that of Yamanaka's, a Harvard/University of California, Los Angeles collaboration, and a group at MIT, published studies that substantially improved on the reprogramming approach, giving rise to iPSCs that were indistinguishable from ESCs. Unlike the first generation of iPSCs, these second generation iPSCs produced viable chimeric mice and contributed to the mouse germline, thereby achieving the 'gold standard' for pluripotent stem cells.

These second-generation iPSCs were derived from mouse fibroblasts by retroviral-mediated expression of the same four transcription factors (Oct4, Sox2, cMyc, Klf4). However, instead of using Fbx15 to select for pluripotent cells, the researchers used Nanog, a gene that is functionally important in ESCs. By using this different strategy, the researchers created iPSCs that were functionally identical to ESCs.

Unfortunately, two of the four genes used (namely, c-Myc and KLF4) are oncogenic, and 20% of the chimeric mice developed cancer. In a later study, Yamanaka reported that one can create iPSCs even without c-Myc. The process takes longer and is not as efficient, but the resulting chimeras didn't develop cancer.

Human Induced Pluripotent Stem Cells

Generation from Human Fibroblasts

In November 2007, a milestone was achieved by creating iPSCs from adult human cells; two independent research teams' studies were released – one in *Science* by James Thomson at University of Wisconsin–Madison and another in *Cell* by Shinya Yamanaka and colleagues at Kyoto University, Japan. With the same principle used earlier in mouse models, Yamanaka had successfully transformed human fibroblasts into pluripotent stem cells using the same four pivotal genes: Oct3/4, Sox2, Klf4, and c-Myc with a retroviral system. Thomson and colleagues used OCT4, SOX2, NANOG, and a different gene LIN28 using a lentiviral system.

Generation from Human Renal Epithelial Cells in Urine

On 8 November 2012, researchers from Austria, Hong Kong and China presented a protocol for generating human iPSCs from exfoliated renal epithelial cells present in urine on Nature Protocols. This method of acquiring donor cells is comparatively less invasive and simple. The team reported the induction procedure to take less time, around 2 weeks for the urinary cell culture and 3 to 4 weeks for the reprogramming; and higher yield, up to 4% using retroviral delivery of exogenous factors. Urinary iPSCs (UiPSCs) were found to show good differentiation potential, and thus represent an alternative choice for producing pluripotent cells from normal individuals or patients with genetic diseases, including those affecting the kidney.

Challenges in Reprogramming Cells to Pluripotency

Although the methods pioneered by Yamanaka and others have demonstrated that adult cells can be reprogrammed to iPS cells, there are still challenges associated with this technology:

1. Low efficiency: in general, the conversion to iPS cells has been incredibly low. For example, the rate at which somatic cells were reprogrammed into iPS cells in Yamanaka's original mouse study was 0.01–0.1%. The low efficiency rate may reflect the need for precise timing, balance, and absolute levels of expression of the reprogramming genes. It may also suggest a need for rare genetic and/or epigenetic changes in the original somatic cell population or in the prolonged culture. However, recently a path was found for efficient reprogramming which required downregulation of the nucleosome remodeling and deacetylation (NuRD) complex. Overexpression of Mbd3, a subunit of NuRD, inhibits induction of iPSCs. Depletion of Mbd3, on the other hand, improves reprogramming efficiency, that results in deterministic and synchronized iPS cell reprogramming (near 100% efficiency within seven days from mouse and human cells).

2. Genomic Insertion: genomic integration of the transcription factors limits the utility of the transcription factor approach because of the risk of mutations being inserted into the target cell's genome. A common strategy for avoiding genomic insertion has been to use a different vector for input. Plasmids, adenoviruses, and transposon vectors have all been explored, but these often come with the tradeoff of lower throughput.

3. Tumors: another main challenge was mentioned above – some of the reprogramming factors are oncogenes that bring on a potential tumor risk. Inactivation or deletion of the tumor suppressor p53, which is the master regulator of cancer, significantly increases reprogramming efficiency. Thus there seems to be a tradeoff between reprogramming efficiency and tumor generation.

4. Incomplete reprogramming: reprogramming also faces the challenge of completeness. This is particularly challenging because the genome-wide epigenetic code must be reformatted to that of the target cell type in order to fully reprogram a cell. However, three separate groups were able to find mouse embryonic fibroblast (MEF)-derived iPS cells that could be injected into tetraploid blastocysts and resulted in the live birth of mice derived entirely from iPS cells, thus ending the debate over the equivalence of embryonic stem cells (ESCs) and iPS with regard to pluripotency.

The table at right summarizes the key strategies and techniques used to develop iPS cells over the past half-decade. Rows of similar colors represents studies that used similar strategies for reprogramming.

Year	Group	Strategy	Contribution
2006	Yamanaka et al.	First to demonstrate	iPS cells were first generated using retroviruses and the four key pluripotency genes; failed to produce viable chimera.
2007	Yamanaka et al.	Different Selection Method	iPS cells were generated again using retroviruses, but this time produced viable chimera (they used different selection methods).
2007	Thomson et al.	Vector	iPS cells were generated again using lentiviruses, and again produced viable chimera
2008	Melton et al.	Small Compound Mimicking	Using HDAC inhibitor valproic acid compensates for C-Myc.
2008	Ding et al.	Small Compound Mimicking	Inhibit HMT with BIX-01294 mimics the effects of Sox2, significantly increases reprogramming efficiency
2008	Hochedlinger et al.	Vector	The group used an adenovirus to avoid the danger of creating tumors, however, this led to lower efficiency
2008	Yamakana et al.	Vector	The group demonstrated reprogramming with no virus (they instead used a plasmid)
2009	Ding et al.	Proteins	Used recombinant proteins ; proteins added to cells via arginine anchors was sufficient to induce pluripotency.
2009	Freed et al.	Vector	Adenoviral gene delivery reprogrammed human fibroblasts to iPS cells.
2009	Blelloch et al.	RNA	Embryonic stem-cell specific microRNAs promted iPS reprogramming
2011	Morrisey et al.	RNA	Demonstrated another method using microRNA that improved the efficency of reprogramming to a rate similar to that demonstrated by Ding.

This timeline summarizes the key strategies and techniques used to develop iPS cells over the past half-decade. Rows of similar colors represents studies that used similar strategies for reprogramming.

Alternative Approaches

Mimicking Transcription Factors with Chemicals

One of the main strategies for avoiding problems (1) and (2) has been to use small compounds that can mimic the effects of transcription factors. These molecule compounds can compensate for a reprogramming factor that does not effectively target the genome or fails at reprogramming for another reason; thus they raise reprogramming efficiency. They also avoid the problem of genomic integration, which in some cases contributes to tumor genesis. Key studies using such strategy were conducted in 2008. Melton et al. studied the effects of histone deacetylase (HDAC) inhibitor valproic acid. They found that it increased reprogramming efficiency 100-fold (compared to Yamanaka's traditional transcription factor method). The researchers proposed that this compound was mimicking the signaling that is usually caused by the transcription factor c-Myc. A similar type of compensation mechanism was proposed to mimic the effects of Sox2. In 2008, Ding et al. used the inhibition of histone methyl transferase (HMT) with BIX-01294 in combination with the activation of calcium channels in the plasma membrane in order to increase reprogramming efficiency. Deng et al. of Beijing University reported on July 2013 that induced pluripotent stem cells can be created without any genetic modification. They used a cocktail of seven small-molecule compounds including DZNep to induce the mouse somatic cells into stem cells which they called CiPS cells with the efficiency – at 0.2% – comparable to those using standard iPSC production techniques. The CiPS cells were introduced into developing mouse embryos and were found to contribute to all major cells types, proving its pluripotency.

Ding et al. demonstrated an alternative to transcription factor reprogramming through the use of drug-like chemicals. By studying the MET (mesenchymal-epithelial transition) process in which fibroblasts are pushed to a stem-cell like state, Ding's group identified two chemicals – ALK5 inhibitor SB431412 and MEK (mitogen-activated protein kinase) inhibitor PD0325901 – which was found to increase the efficiency of the classical genetic method by 100 fold. Adding a third compound known to be involved in the cell survival pathway, Thiazovivin further increases the efficiency by 200 fold. Using the combination of these three compounds also decreased the reprogramming process of the human fibroblasts from four weeks to two weeks.

In April 2009, it was demonstrated that generation of iPS cells is possible without any genetic alteration of the adult cell: a repeated treatment of the cells with certain proteins channeled into the cells via poly-arginine anchors was sufficient to induce pluripotency. The acronym given for those iPSCs is piPSCs (protein-induced pluripotent stem cells).

Alternate Vectors

Another key strategy for avoiding problems such as tumor genesis and low throughput has been to use alternate forms of vectors: adenovirus, plasmids, and naked DNA and/ or protein compounds.

In 2008, Hochedlinger et al. used an adenovirus to transport the requisite four transcription factors into the DNA of skin and liver cells of mice, resulting in cells identical to ESCs. The adenovirus is unique from other vectors like viruses and retroviruses because it does not incorporate any of its own genes into the targeted host and avoids the potential for insertional mutagenesis. In 2009, Freed et al. demonstrated successful reprogramming of human fibroblasts to iPS cells. Another advantage of using adenoviruses is that they only need to present for a brief amount of time in order for effective reprogramming to take place.

Also in 2008, Yamanaka et al. found that they could transfer the four necessary genes with a plasmid. The Yamanaka group successfully reprogrammed mouse cells by transfection with two plasmid constructs carrying the reprogramming factors; the first plasmid expressed c-Myc, while the second expressed the other three factors (Oct4, Klf4, and Sox2). Although the plasmid methods avoid viruses, they still require cancer-promoting genes to accomplish reprogramming. The other main issue with these methods is that they tend to be much less efficient compared to retroviral methods. Furthermore, transfected plasmids have been shown to integrate into the host genome and therefore they still pose the risk of insertional mutagenesis. Because non-retroviral approaches have demonstrated such low efficiency levels, researchers have attempted to effectively rescue the technique with what is known as the PiggyBac Transposon System. The lifecycle of this system is shown below. Several studies have demonstrated that this system can effectively deliver the key reprogramming factors without leaving any footprint mutations in the host cell genome. As demonstrated in the figure, the piggyBac transposon system involves the re-excision of exogenous genes, which eliminates issues like insertional mutagenesis

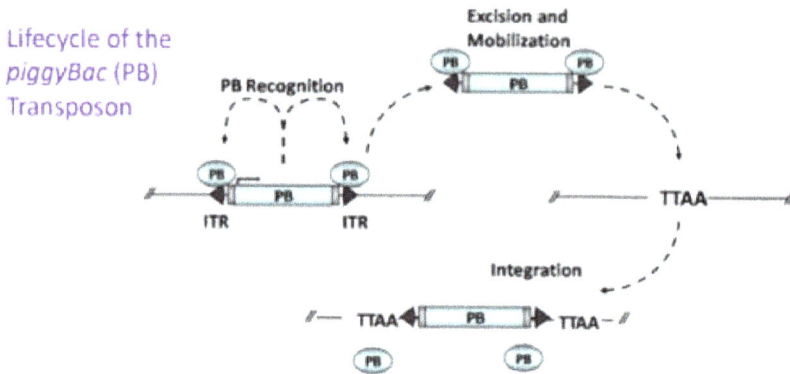

Lifecycle of the *piggyBac* (PB) Transposon

Lifecycle of the PiggyBac Transposon System

Stimulus-triggered Acquisition of Pluripotency Cell

In January 2014, two articles were published claiming that a type of pluripotent stem cell can be generated by subjecting the cells to certain types of stress (bacterial toxin, a low pH of 5.7, or physical squeezing); the resulting cells were called STAP cells, for stimulus-triggered acquisition of pluripotency.

In light of difficulties that other labs had replicating the results of the surprising study, in March 2014, one of the co-authors has called for the articles to be retracted. On 4 June 2014, the lead author, Obokata agreed to retract both the papers after she was found to have committed 'research misconduct' as concluded in an investigation by RIKEN on 1 April 2014.

RNA Molecules

Studies by Blelloch et al. in 2009 demonstrated that expression of ES cell-specific microRNA molecules (such as miR-291, miR-294 and miR-295) enhances the efficiency of induced pluripotency by acting downstream of c-Myc . More recently (in April 2011), Morrisey et al. demonstrated another method using microRNA that improved the efficiency of reprogramming to a rate similar to that demonstrated by Ding. MicroRNAs are short RNA molecules that bind to complementary sequences on messenger RNA and block expression of a gene. Morrisey's team worked on microRNAs in lung development, and hypothesized that their microRNAs perhaps blocked expression of repressors of Yamanaka's four transcription factors. Possible mechanisms by which microRNAs can induce reprogramming even in the absence of added exogenous transcription factors, and how variations in microRNA expression of iPS cells can predict their differentiation potential discussed by Xichen Bao et al.

Transcription Factors of Induction

The generation of iPS cells is crucially dependent on the transcription factors used for the induction.

Oct-3/4 and certain products of the Sox gene family (Sox1, Sox2, Sox3, and Sox15) have been identified as crucial transcriptional regulators involved in the induction process whose absence makes induction impossible. Additional genes, however, including certain members of the Klf family (Klf1, Klf2, Klf4, and Klf5), the Myc family (c-myc, L-myc, and N-myc), Nanog, and LIN28, have been identified to increase the induction efficiency.

- Oct-3/4 (Pou5f1) Oct-3/4 is one of the family of octamer ("Oct") transcription factors, and plays a crucial role in maintaining pluripotency. The absence of Oct-3/4 in Oct-3/4$^+$ cells, such as blastomeres and embryonic stem cells, leads to spontaneous trophoblast differentiation, and presence of Oct-3/4 thus gives rise to the pluripotency and differentiation potential of embryonic stem cells. Various other genes in the "Oct" family, including Oct-3/4's close relatives, Oct1 and Oct6, fail to elicit induction, thus demonstrating the exclusiveness of Oct-3/4 to the induction process.

- Sox family: The Sox family of transcription factors is associated with maintaining pluripotency similar to Oct-3/4, although it is associated with multipotent and unipotent stem cells in contrast with Oct-3/4, which is exclusively expressed in pluripotent stem cells. While Sox2 was the initial gene used for induction by Yamanaka et al., Jaenisch et al., and Thomson et al., other transcription factors in the Sox family have been found to work as well in the induction process. Sox1 yields iPS cells with a similar efficiency as Sox2, and genes Sox3, Sox15, and Sox18 also generate iPS cells, although with decreased efficiency.

- Klf family: Klf4 of the Klf family of transcription factors was initially identified by Yamanaka et al. and confirmed by Jaenisch et al. as a factor for the generation of mouse iPS cells and was demonstrated by Yamanaka et al. as a factor for generation of human iPS cells. However, Thomson et al. reported that Klf4 was unnecessary for generation of human iPS cells and in fact failed to generate human iPS cells. Klf2 and Klf4 were found to be factors capable of generating iPS cells, and related genes Klf1 and Klf5 did as well, although with reduced efficiency.

- Myc family: The Myc family of transcription factors are proto-oncogenes implicated in cancer. Yamanaka et al. and Jaenisch et al. demonstrated that c-myc is a factor implicated in the generation of mouse iPS cells and Yamanaka et al. demonstrated it was a factor implicated in the generation of human iPS cells. However, Thomson et al., Yamanaka et al. Usage of the "myc" family of genes in induction of iPS cells is troubling for the eventuality of iPS cells as clinical therapies, as 25% of mice transplanted with c-myc-induced iPS cells developed lethal teratomas. N-myc and L-myc have been identified to induce instead of c-myc with similar efficiency.

- Nanog: In embryonic stem cells, Nanog, along with Oct-3/4 and Sox2, is necessary in promoting pluripotency. Therefore, it was surprising when Yamanaka et al. reported that Nanog was unnecessary for induction although Thomson et al. has reported it is possible to generate iPS cells with Nanog as one of the factors.

- LIN28: LIN28 is an mRNA binding protein expressed in embryonic stem cells and embryonic carcinoma cells associated with differentiation and proliferation. Thomson et al. demonstrated it is a factor in iPSC generation, although it is unnecessary.

- Glis1: Glis1 is transcription factor that can be used with Oct-3/4, Sox2 and Klf4 to induce pluripotency. It poses numerous advantages when used instead of C-myc.

Identity

Induced pluripotent stem cells are similar to natural pluripotent stem cells, such as embryonic stem (ES) cells, in many aspects, such as the expression of certain stem cell genes and proteins, chromatin methylation patterns, doubling time, embryoid body formation, teratoma formation, viable chimera formation, and potency and differentiability, but the full extent of their relation to natural pluripotent stem cells is still being assessed.

Gene expression and genome-wide H3K4me3 and H3K27me3 were found to be extremely similar between ES and iPS cells. The generated iPSCs were remarkably similar to naturally isolated pluripotent stem cells (such as mouse and human embryonic stem cells, mESCs and hESCs, respectively) in the following respects, thus confirming the identity, authenticity, and pluripotency of iPSCs to naturally isolated pluripotent stem cells:

- Cellular biological properties

  o Morphology: iPSCs were morphologically similar to ESCs. Each cell had round shape, large nucleolus and scant cytoplasm. Colonies of iPSCs were also similar to that of ESCs. Human iPSCs formed sharp-edged, flat, tightly packed colonies similar to hESCs and mouse iPSCs formed the colonies similar to mESCs, less flat and more aggregated colonies than that of hESCs.

  o Growth properties: Doubling time and mitotic activity are cornerstones of ESCs, as stem cells must self-renew as part of their definition. iPSCs were mitotically active, actively self-renewing, proliferating, and dividing at a rate equal to ESCs.

  o Stem cell markers: iPSCs expressed cell surface antigenic markers expressed on ESCs. Human iPSCs expressed the markers specific to hESC, including SSEA-3, SSEA-4, TRA-1-60, TRA-1-81, TRA-2-49/6E, and Nanog. Mouse iPSCs expressed SSEA-1 but not SSEA-3 nor SSEA-4, similarly to mESCs.

o   Stem Cell Genes: iPSCs expressed genes expressed in undifferentiated ESCs, including Oct-3/4, Sox2, Nanog, GDF3, REX1, FGF4, ESG1, DPPA2, DPPA4, and hTERT.

o   Telomerase activity: Telomerases are necessary to sustain cell division unrestricted by the Hayflick limit of ~50 cell divisions. hESCs express high telomerase activity to sustain self-renewal and proliferation, and iPSCs also demonstrate high telomerase activity and express hTERT (human telomerase reverse transcriptase), a necessary component in the telomerase protein complex.

- Pluripotency: iPSCs were capable of differentiation in a fashion similar to ESCs into fully differentiated tissues.

o   Neural differentiation: iPSCs were differentiated into neurons, expressing βIII-tubulin, tyrosine hydroxylase, AADC, DAT, ChAT, LMX1B, and MAP2. The presence of catecholamine-associated enzymes may indicate that iPSCs, like hESCs, may be differentiable into dopaminergic neurons. Stem cell-associated genes were downregulated after differentiation.

o   Cardiac differentiation: iPSCs were differentiated into cardiomyocytes that spontaneously began beating. Cardiomyocytes expressed TnTc, MEF2C, MYL2A, MYHCβ, and NKX2.5. Stem cell-associated genes were downregulated after differentiation.

o   Teratoma formation: iPSCs injected into immunodeficient mice spontaneously formed teratomas after nine weeks. Teratomas are tumors of multiple lineages containing tissue derived from the three germ layers endoderm, mesoderm and ectoderm; this is unlike other tumors, which typically are of only one cell type. Teratoma formation is a landmark test for pluripotency.

o   Embryoid body: hESCs in culture spontaneously form ball-like embryo-like structures termed "embryoid bodies", which consist of a core of mitotically active and differentiating hESCs and a periphery of fully differentiated cells from all three germ layers. iPSCs also form embryoid bodies and have peripheral differentiated cells.

o   Chimeric mice: hESCs naturally reside within the inner cell mass (embryoblast) of blastocysts, and in the embryoblast, differentiate into the embryo while the blastocyst's shell (trophoblast) differentiates into extraembryonic tissues. The hollow trophoblast is unable to form a living embryo, and thus it is necessary for the embryonic stem cells within the embryoblast to differentiate and form the embryo. iPSCs were injected by micropipette into a trophoblast, and the blastocyst was transferred to recipient females. Chimeric living mouse pups were created: mice with iPSC derivatives incorporated all across their bodies with 10%–90% chimerism.

   o Tetraploid complementation: iPS cells from mouse fetal fibroblasts injected into tetraploid blastocysts (which themselves can only form extra-embryonic tissues) can form whole, non-chimeric, fertile mice, although with low success rate.

- Epigenetic reprogramming

   o Promoter demethylation: Methylation is the transfer of a methyl group to a DNA base, typically the transfer of a methyl group to a cytosine molecule in a CpG site (adjacent cytosine/guanine sequence). Widespread methylation of a gene interferes with expression by preventing the activity of expression proteins, or by recruiting enzymes that interfere with expression. Thus, methylation of a gene effectively silences it by preventing transcription. Promoters of pluripotency-associated genes, including Oct-3/4, Rex1, and Nanog, were demethylated in iPSCs, demonstrating their promoter activity and the active promotion and expression of pluripotency-associated genes in iPSCs.

   o DNA methylation globally: Human iPS cells are highly similar to ES cells in their patterns of which cytosines are methylated, more than to any other cell type. However, on the order of a thousand sites show differences in several iPS cell lines. Half of these resemble the somatic cell line the iPS cells were derived from, the rest are iPSC-specific. Tens of regions which are megabases in size have also been found where iPS cells are not reprogrammed to the ES cell state.

   o Histone demethylation: Histones are compacting proteins that are structurally localized to DNA sequences that can affect their activity through various chromatin-related modifications. H3 histones associated with Oct-3/4, Sox2, and Nanog were demethylated, indicating the expression of Oct-3/4, Sox2, and Nanog.

Safety

- The major concern with the potential clinical application of iPSCs is their propensity to form tumors. Much the same as ESC, iPSCs readily form teratoma when injected into immunodeficient mice. Teratoma formation is considered a major obstacle to stem-cell based regenerative medicine by the FDA.

- A more recent study on motor functional recovery after spinal cord injuries in mice showed that after human-induced pluripotent stem cells were transplanted into the mice, the cells differentiated into three neural lineages in the spinal cord. The cells stimulated regrowth of the damaged spinal cord, maintained myelination, and formed synapses. These positive outcomes were observed for over 112 days after the spinal cord injury, without tumor formation. Nevertheless, a follow-up study by the same group showed distinct clones of human-induced pluripotent stem cells eventually formed tumors.

- Since iPSCs can only be produced with high efficiency at this time using modifications, they are generally predicted to be less safe and more tumorigenic than hESC. All the genes that have been shown to promote iPSC formation have also been linked to cancer in one way or another. Some of the genes are known oncogenes, including the members of the Myc family. While omitting Myc allows for IPSC formation, the efficiency is reduced up to 100 fold.

- A non-genetic method of producing iPSCs has been demonstrated using recombinant proteins, but its efficiency was quite low. However, refinements to this methodology yielding higher efficiency may lead to production of safer iPSCs. Other approaches such as using adenovirus or plasmids are generally thought to be safer than retroviral methods.

- An important area for future studies in the iPSC field is directly testing iPSC tumorigenicity using methods that mimic the approaches that would be used for regenerative medicine therapies. Such studies are crucial since iPSCs not only form teratoma, but also mice derived from iPSCs have a high incidence of death from malignant cancer. A 2010 paper was published in the journal Stem Cells indicating that iPS cells are far more tumorigenic than ESC, supporting the notion that iPS cell safety is a serious concern.

- Concern regarding the immunogenicity of IPS cells arose in 2011 when Zhou et al. performed a study involving a teratomaformation assay and demonstrated that IPS cells produced an immune response strong enough to cause rejection of the cells. When a similar procedure was performed on genetically equivalent ES cells however, Zhou et al. found teratomas, which indicated that the cells were tolerated by the immune system. In 2013, Araki et al. attempted to reproduce the conclusion obtained by Zhou et al. using a different procedure. They took cells from a chimera that had been grown from IPSC clones and a mouse embryo, this tissue was then transplanted into syngenic mice. They conducted a similar trial using ES cells instead of IPSC clone and compared the results. Findings indicate that there was no significant difference in the immunogenic response produced by the IPS cells and the ES cells. Furthermore, Araki et al. reported little or no immunogenic response for both cell lines. Thus, Araki et al. was unable to come to the same conclusion as Zhou et al.

Recent achievements and future tasks for safe iPSC-based cell therapy are collected in the review of Okano et al.

Medical Research

The task of producing iPS cells continues to be challenging due to the six problems mentioned above. A key tradeoff to overcome is that between efficiency and genomic integration. Most methods that do not rely on the integration of transgenes are inefficient, while those that do rely on the integration of transgenes face the problems of incomplete reprogramming and tumor genesis, although a vast number of techniques and methods have been attempted. Another large set of strategies is to perform a proteomic characterization of iPS cells. The Wu group at Stanford University has made significant progress with this strategy. Further studies and new strategies should generate optimal solutions to the five main challenges. One approach might attempt to combine the positive attributes of these strategies into an ultimately effective technique for reprogramming cells to iPS cells.

Another approach is the use of iPS cells derived from patients to identify therapeutic drugs able to rescue a phenotype. For instance, iPS cell lines derived from patients affected by ectodermal dysplasia syndrome (EEC), in which the p63 gene is mutated, display abnormal epithelial commitment that could be partially rescued by a small compound

Disease Modeling and Drug Development

An attractive feature of human iPS cells is the ability to derive them from adult patients to study the cellular basis of human disease. Since iPS cells are self-renewing and pluripotent, they represent a theoretically unlimited source of patient-derived cells which can be turned into any type of cell in the body. This is particularly important because many other types of human cells derived from patients tend to stop growing after a few passages in laboratory culture. iPS cells have been generated for a wide variety of human genetic diseases, including common disorders such as Down syndrome and polycystic kidney disease. In many instances, the patient-derived iPS cells exhibit cellular defects not observed in iPS cells from healthy patients, providing insight into the pathophysiology of the disease. An international collaborated project, StemBANCC, was formed in 2012 to build a collection of iPS cell lines for drug screening for a variety of disease. Managed by the University of Oxford, the effort pooled funds and resources from 10 pharmaceutical companies and 23 universities. The goal is to generate a library of 1,500 iPS cell lines which will be used in early drug testing by providing a simulated human disease environment. Furthermore, combining hiPSC technology and genetically-encoded voltage and calcium indicators provided a large-scale and high-throughput platform for cardiovascular drug safety screening.

Organ Synthesis

A proof-of-concept of using induced pluripotent stem cells (iPSCs) to generate human organ for transplantation was reported by researchers from Japan. Human 'liver buds' (iPSC-LBs) were grown from a mixture of three different kinds of stem cells: hepatocytes (for liver function) coaxed from iPSCs; endothelial stem cells (to form lining of blood vessels) from umbilical cord blood; and mesenchymal stem cells (to form connective tissue). This new approach allows different cell types to self-organize into a complex organ, mimicking the process in fetal development. After growing *in vitro* for a few days, the liver buds were transplanted into mice where the 'liver' quickly connected with the host blood vessels and continued to grow. Most importantly, it performed regular liver functions including metabolizing drugs and producing liver-specific proteins. Further studies will monitor the longevity of the transplanted organ in the host body (ability to integrate or avoid rejection) and whether it will transform into tumors. Using this method, cells from one mouse could be used to test 1,000 drug compounds to treat liver disease, and reduce animal use by up to 50,000.

Tissue Repair

Embryonic cord-blood cells were induced into pluripotent stem cells using plasmid DNA. Using cell surface endothelial/pericytic markers CD31 and CD146, researchers identified 'vascular progenitor', the high-quality, multipotent vascular stem cells. After the iPS cells were injected directly into the vitreous of the damaged retina of mice, the stem cells engrafted into the retina, grew and repaired the vascular vessels.

Labelled iPSCs-derived NSCs injected into laboratory animals with brain lesions were shown to migrate to the lesions and some motor function improvement was observed.

Red Blood Cells

Although a pint of donated blood contains about two trillion red blood cells and over 107 million blood donations are collected globally, there is still a critical need for blood for transfusion. In 2014, type O red blood cells were synthesized at the Scottish National Blood Transfusion Service from iPSC. The cells were induced to become a mesoderm and then blood cells and then red blood cells. The final step was to make them eject their nuclei and mature properly. Type O can be transfused into all patients. Human clinical trials were not expected to begin before 2016.

Clinical Trial

The first human clinical trial using autologous iPSCs was approved by the Japan Ministry Health and was to be conducted in 2014 in Kobe. However the trial was suspended after Japan's new regenerative medicine laws came into effect last November. iPSCs derived from skin cells from six patients suffering from wet age-related macular degeneration were to be reprogrammed to differentiate into retinal pigment epithelial (RPE) cells. The cell sheet would be transplanted into the affected retina where the degenerated RPE tissue was excised. Safety and vision restoration monitoring would last one to three years. The benefits of using autologous iPSCs are that there is theoretically no risk of rejection and it eliminates the need to use embryonic stem cells.

Induced Stem Cells

Induced stem cells (iSC) are stem cells derived from somatic, reproductive, pluripotent or other cell types by deliberate epigenetic reprogramming. They are classified as either totipotent (iTC), pluripotent (iPSC) or progenitor (multipotent—iMSC, also called an induced multipotent progenitor cell—iMPC) or unipotent—(iUSC) according to their developmental potential and degree of dedifferentiation. Progenitors are obtained by so-called direct reprogramming or directed differentiation and are also called induced somatic stem cells.

Three techniques are widely recognized:

- Transplantation of nuclei taken from somatic cells into an oocyte (egg cell) lacking its own nucleus (removed in lab)

- Fusion of somatic cells with pluripotent stem cells and

- Transformation of somatic cells into stem cells, using the genetic material encoding reprogramming protein factors, recombinant proteins; microRNA, a synthetic, self-replicating polycistronic RNA and low-molecular weight biologically active substances.

Natural Processes

In 1895 Thomas Morgan removed one of a frog's two blastomeres and found that amphibians are able to form whole embryos from the remaining part. This meant that the cells can change their differentiation pathway. In 1924 Spemann and Mangold demonstrated the key importance of cell–cell inductions during animal development. The reversible transformation of cells of one differentiated cell type to another is called metaplasia. This transition can be a part of the normal maturation process, or caused by an inducement.

One example is the transformation of iris cells to lens cells in the process of maturation and transformation of retinal pigment epithelium cells into the neural retina during regeneration in adult newt eyes. This process allows the body to replace cells not suitable to new conditions with more suitable new cells. In Drosophila imaginal discs, cells have to choose from a limited number of standard discrete differentiation states. The fact that transdetermination (change of the path of differentiation) often occurs for a group of cells rather than single cells shows that it is induced rather than part of maturation.

The researchers were able to identify the minimal conditions and factors that would be sufficient for starting the cascade of molecular and cellular processes to instruct pluripotent cells to organize the embryo. They showed that opposing gradients of bone morphogenetic protein (BMP) and Nodal, two transforming growth factor family members that act as morphogens, are sufficient to induce molecular and cellular mechanisms required to organize, *in vivo* or *in vitro*, uncommitted cells of the zebrafish blastula animal pole into a well-developed embryo.

Some types of mature, specialized adult cells can naturally revert to stem cells. For example, "chief" cells express the stem cell marker Troy. While they normally produce digestive fluids for the stomach, they can revert into stem cells to make temporary repairs to stomach injuries, such as a cut or damage from infection. Moreover, they can make this transition even in the absence of noticeable injuries and are capable of replenishing entire gastric units, in essence serving as quiescent "reserve" stem cells. Differentiated airway epithelial cells can revert into stable and functional stem cells in vivo.

After injury, mature terminally differentiated kidney cells dedifferentiate into more primordial versions of themselves and then differentiate into the cell types needing replacement in the damaged tissue Macrophages can self-renew by local proliferation of mature differentiated cells. In newts, muscle tissue is regenerated from specialized muscle cells that dedifferentiate and forget the type of cell they had been. This capacity to regenerate does not decline with age and may be linked to their ability to make new stem cells from muscle cells on demand.

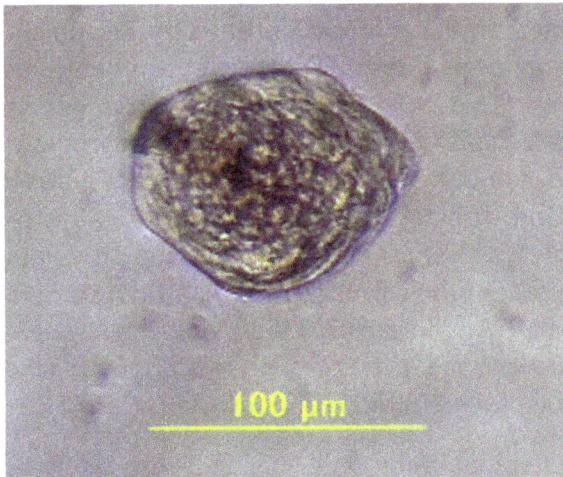

Cluster forming of pluripotent Muse/Stem cell

A variety of nontumorigenic stem cells display the ability to generate multiple cell types. For instance, multilineage-differentiating stress-enduring (Muse) cells are stress-tolerant adult human stem cells that can self-renew. They form characteristic cell clusters in suspension culture that express a set of genes associated with pluripotency and can differentiate into endodermal, ectodermal and mesodermal cells both in vitro and in vivo.

Other well-documented examples of transdifferentiation and their significance in development and regeneration were described in detail.

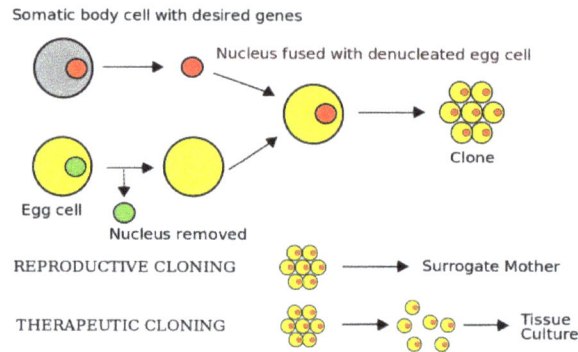

Induced totipotent cells usually can be obtained by reprogramming somatic cells by somatic-cell nuclear transfer (SCNT).

Induced Totipotent Cells

SCNT-mediated

Induced totipotent cells can be obtained by reprogramming somatic cells with somatic-cell nuclear transfer (SCNT). The process involves sucking out the nucleus of a somatic (body) cell and injecting it into an oocyte that has had its nucleus removed

Using an approach based on the protocol outlined by Tachibana et al., hESCs can be generated by SCNT using dermal fibroblasts nuclei from both a middle-aged 35-year-old male and an elderly, 75-year-old male, suggesting that age-associated changes are not necessarily an impediment to SCNT-based nuclear reprogramming of human cells. Such reprogramming of somatic cells to a pluripotent state holds huge potentials for regenerative medicine. Unfortunately, the cells generated by this technology, potentially are not completely protected from the immune system of the patient (donor of nuclei), because they have the same mitochondrial DNA, as a donor of oocytes, instead of the patients mitochondrial DNA. This reduces their value as a source for autologous stem cell transplantation therapy, as for the present, it is not clear whether it can induce an immune response of the patient upon treatment.

Induced androgenetic haploid embryonic stem cells can be used instead of sperm for cloning. These cells, synchronized in M phase and injected into the oocyte can produce viable offspring.

These developments, together with data on the possibility of unlimited oocytes from mitotically active reproductive stem cells, offer the possibility of industrial production of transgenic farm animals. Repeated recloning of viable mice through a SCNT method that includes a histone deacetylase inhibitor, trichostatin, added to the cell culture medium, show that it may be possible to reclone animals indefinitely with no visible accumulation of reprogramming or genomic errors However, research into technologies to develop sperm and egg cells from stem cells raises bioethical issues.

Such technologies may also have far-reaching clinical applications for overcoming cytoplasmic defects in human oocytes. For example, the technology could prevent inherited mitochondrial disease from passing to future generations. Mitochondrial genetic material is passed from mother to child. Mutations can cause diabetes, deafness, eye disorders, gastrointestinal disorders, heart disease, dementia and other neurological diseases. The nucleus from one human egg has been transferred to another, including its mitochondria, creating a cell that could be regarded as having two mothers. The eggs were then fertilised and the resulting embryonic stem cells carried the swapped mitochondrial DNA. As evidence that the technique is safe author of this method points to the existence of the healthy monkeys that are now more than four years old — and are the product of mitochondrial transplants across different genetic backgrounds.

In late-generation telomerase-deficient (Terc$-/-$) mice, SCNT-mediated reprogramming mitigates telomere dysfunction and mitochondrial defects to a greater extent than iPSC-based reprogramming.

Other cloning and totipotent transformation achievements have been described.

Obtained Without SCNT

Recently some researchers succeeded to get the totipotent cells without the aid of SCNT. Totipotent cells were obtained using the epigenetic factors such as oocyte germinal isoform of histone. Reprogramming in vivo, by transitory induction of the four factors Oct4, Sox2, Klf4 and c-Myc in mice, confers totipotency features. Intraperitoneal injection of such in vivo iPS cells generates embryo-like structures that express embryonic and extraembryonic (trophectodermal) markers.

Rejuvenation to iPSc

iPSc were first obtained in the form of transplantable teratocarcinoma induced by grafts taken from mouse embryos. Teratocarcinoma formed from somatic cells. Genetically mosaic mice were obtained from malignant teratocarcinoma cells, confirming the cells' pluripotency. It turned out that teratocarcinoma cells are able to maintain a culture of pluripotent embryonic stem cell in an undifferentiated state, by supplying the culture medium with various factors. In the 1980s, it became clear that transplanting pluripotent/embryonic stem cells into the body of adult mammals, usually leads to the formation of teratomas, which can then turn into a malignant tumor teratocarcinoma. However, putting teratocarcinoma cells into the embryo at the blastocyst stage, caused them to become incorporated in the inner cell mass and often produced a normal chimeric (i.e. composed of cells from different organisms) animal. This indicated that the cause of the teratoma is a dissonance - mutual miscommunication between young donor cells and surrounding adult cells (the recipient's so-called "niche").

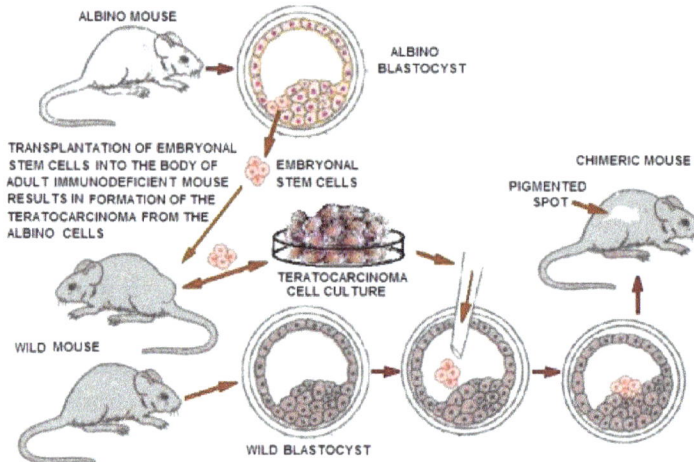

Transplantation of pluripotent/embryonic stem cells into the body of adult mammals, usually leads to the formation of teratomas, which can then turn into a malignant tumor teratocarcinoma. However, putting teratocarcinoma cells into the embryo at the blastocyst stage, caused them to become incorporated in the cell mass and often produced a normal healthy chimeric (i.e. composed of cells from different organisms) animal

In August 2006, Japanese researchers circumvented the need for an oocyte, as in SCNT. By reprograming mouse embryonic fibroblasts into pluripotent stem cells via the ectopic expression of four transcription factors, namely Oct4, Sox2, Klf4 and c-Myc, they proved that the overexpression of a small number of factors can push the cell to transition to a new stable state that is associated with changes in the activity of thousands of genes.

Reprogramming mechanisms are thus linked, rather than independent and are centered on a small number of genes. IPSC properties are very similar to ESCs. iPSCs have been shown to support the development of all-iPSC mice using a tetraploid (4n) embryo, the most stringent assay for developmental potential. However, some genetically normal iPSCs failed to produce all-iPSC mice because of aberrant epigenetic silencing of the imprinted Dlk1-Dio3 gene cluster.

An important advantage of iPSC over ESC is that they can be derived from adult cells, rather than from embryos. Therefore, it became possible to obtain iPSC from adult and even elderly patients.

Reprogramming somatic cells to iPSC leads to rejuvenation. It was found that reprogramming leads to telomere lengthening and subsequent shortening after their differentiation back into fibroblast-like derivatives. Thus, reprogramming leads to the restoration of embryonic telomere length, and hence increases the potential number of cell divisions otherwise limited by the Hayflick limit.

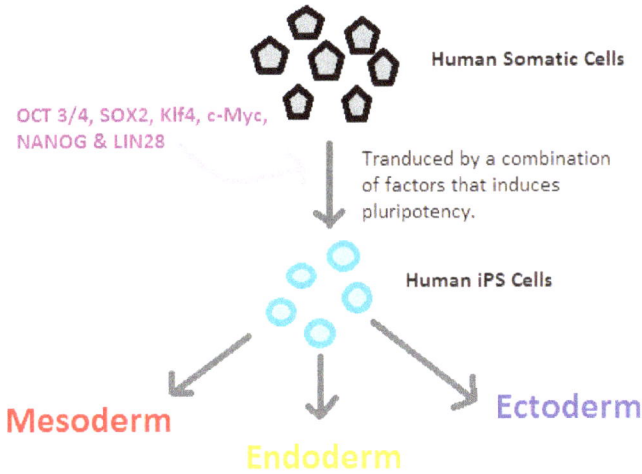

Human somatic cells are made pluripotent by transducing them with factors that induces pluripotency (OCT 3/4, SOX2, Klf4, c-Myc, NANOG and LIN28). This results in the production of IPS cells, which can differentiate into any cells of the three embryonic germ layers (Mesoderm, Endoderm, Ectoderm).

However, because of the dissonance between rejuvenated cells and the surrounding niche of the recipient's older cells, the injection of his own iPSC usually leads to an immune response, which can be used for medical purposes, or the formation of tumors such as teratoma. The reason has been hypothesized to be that some cells differentiated from ESC and iPSC in vivo continue to synthesize embryonic protein isoforms. So, the immune system might detect and attack cells that are not cooperating properly.

A small molecule called MitoBloCK-6 can force the pluripotent stem cells to die by triggering apoptosis (via cytochrome c release across the mitochondrial outer membrane) in human pluripotent stem cells, but not in differentiated cells. Shortly after differentiation, daughter cells became resistant to death. When MitoBloCK-6 was introduced to differentiated cell lines, the cells remained healthy. The key to their survival, was hypothesized to be due to the changes undergone by pluripotent stem cell mitochondria in the process of cell differentiation. This ability of MitoBloCK-6 to separate the pluripotent and differentiated cell lines has the potential to reduce the risk of teratomas and other problems in regenerative medicine.

In 2012 other small molecules (selective cytotoxic inhibitors of human pluripotent stem cells—hPSCs) were identified that prevented human pluripotent stem cells from forming teratomas in mice. The most potent and selective compound of them (PluriSIn #1) inhibits stearoyl-coA desaturase (the key enzyme in oleic acid biosynthesis), which finally results in apoptosis. With the help of this molecule the undifferentiated cells can be selectively removed from culture. An efficient strategy to selectively eliminate pluripotent cells with teratoma potential is targeting pluripotent stem cell-specific antiapoptotic factor(s) (i.e., survivin or Bcl10). A single treatment with chemical survivin inhibitors (e.g., quercetin or YM155) can induce selective and complete cell death of undifferentiated hPSCs and is claimed to be sufficient to prevent teratoma formation after transplantation. However, it is unlikely that any kind of preliminary clearance, is able to secure the replanting iPSC or ESC. After the selective removal of pluripotent cells, they re-emerge quickly by reverting differentiated cells into stem cells, which leads to tumors. This may be due to the disorder of let-7 regulation of its target Nr6a1 (also known as Germ cell nuclear factor - GCNF), an embryonic transcriptional repressor of pluripotency genes that regulates gene expression in adult fibroblasts following micro-RNA miRNA loss.

Teratoma formation by pluripotent stem cells may be caused by low activity of PTEN enzyme, reported to promote the survival of a small population (0,1-5% of total population) of highly tumorigenic, aggressive, teratoma-initiating embryonic-like carcinoma cells during differentiation. The survival of these teratoma-initiating cells is associated with failed repression of Nanog as well as a propensity for increased glucose and cholesterol metabolism. These teratoma-initiating cells also expressed a lower ratio of p53/p21 when compared to non-tumorigenic cells. In connection with the above safety problems, the use iPSC for cell therapy is still limited. However, they can be used for a variety of other purposes - including the modeling of disease, screening (selective selection) of drugs, toxicity testing of various drugs.

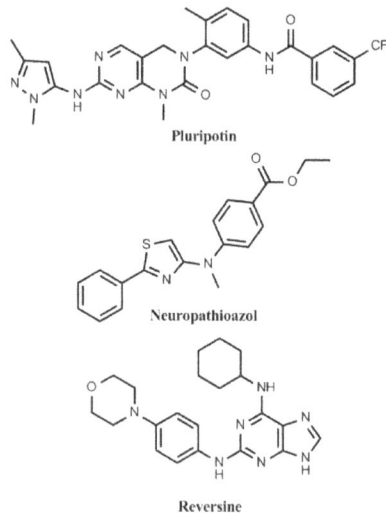

Pluripotin

Neuropathioazol

Reversine

Small molecule modulators of stem-cell fate.

It is interesting to note that the tissue grown from iPSCs, placed in the "chimeric" embryos in the early stages of mouse development, practically do not cause an immune response (after the embryos have grown into adult mice) and are suitable for autologous transplantation At the same time, full reprogramming of adult cells in vivo within tissues by transitory induction of the four factors Oct4, Sox2, Klf4 and c-Myc in mice results in teratomas emerging from multiple organs. Furthermore, partial reprogramming of cells toward pluripotency in vivo in mice demonstrates that incomplete reprogramming entails epigenetic changes (failed repression of Polycomb targets and altered DNA methylation) in cells that drive cancer development.

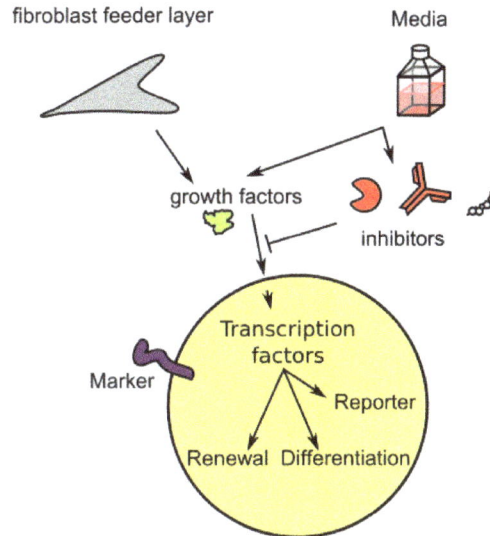

Cell culture example of a small molecule as a tool instead of a protein. in cell culture to obtain a pancreatic lineage from mesodermal stem cells the retinoic acid signalling pathway must be activated while the sonic hedgehog pathway inhibited, which can be done by adding to the media anti-shh antibodies, Hedgehog interacting protein or cyclopamine, the first two are protein and the last a small molecule.

Mogrify Algorithm

Determining the unique set of cellular factors that is needed to be manipulated for each cell conversion is a long and costly process that involved much trial and error. As a result, this first step of identifying the key set of cellular factors for cell conversion is the major obstacle researchers face in the field of cell reprogramming. An international team of researchers have developed an algorithm, called Mogrify(1), that can predict the optimal set of cellular factors required to convert one human cell type to another. When tested, Mogrify was able to accurately predict the set of cellular factors required for previously published cell conversions correctly. To further validate Mogrify's predictive ability, the team conducted two novel cell conversions in the laboratory using human cells and these were successful in both attempts solely using the predictions of Mogrify. Mogrify has been made available online for other researchers and scientists.

Chemical Inducement

By using solely small molecules, Deng Hongkui and colleagues demonstrated that endogenous "master genes" are enough for cell fate reprogramming. They induced a pluripotent state in adult cells from mice using seven small-molecule compounds. The effectiveness of the method is quite high: it was able to convert 0.02% of the adult tissue cells into iPSCs, which is comparable to the gene insertion conversion rate. The authors note that the mice generated from CiPSCs were "100% viable and apparently healthy for up to 6 months". So, this chemical reprogramming strategy has potential use in generating functional desirable cell types for clinical applications.

In 2015th year a robust chemical reprogramming system was established with a yield up to 1,000-fold greater than that of the previously reported protocol. So, chemical reprogramming became a promising approach to manipulate cell fates.

Differentiation from Induced Teratoma

The fact that human iPSCs capable of forming teratomas not only in humans but also in some animal body, in particular in mice or pigs, allowed to develop a method for differentiation of iPSCs in vivo. For this purpose, iPSCs with an agent for inducing differentiation into target cells are injected to genetically modified pig or mouse that has suppressed immune system activation on human cells. The formed teratoma is cut out and used for the isolation of the necessary differentiated human cells by means of monoclonal antibody to tissue-specific markers on the surface of these cells. This method has been successfully used for the production of functional myeloid, erythroid and lymphoid human cells suitable for transplantation (yet only to mice). Mice engrafted with human iPSC teratoma-derived hematopoietic cells produced human B and T cells capable of functional immune responses. These results offer hope that in vivo generation of patient customized cells is feasible, providing materials that could be useful for transplantation, human antibody generation and drug screening applications. Using MitoBloCK-6 and/or PluriSIn # 1 the differentiated progenitor cells can be further purified from teratoma forming pluripotent cells. The fact, that the differentiation takes place even in the teratoma niche, offers hope that the resulting cells are sufficiently stable to stimuli able to cause their transition back to the dedifferentiated (pluripotent) state and therefore safe. A similar in vivo differentiation system, yielding engraftable hematopoietic stem cells from mouse and human iPSCs in teratoma-bearing animals in combination with a maneuver to facilitate hematopoiesis, was described by Suzuki et al. They noted that neither leukemia nor tumors were observed in recipients after intravenous injection of iPSC-derived hematopoietic stem cells into irradiated recipients. Moreover, this injection resulted in multilineage and long-term reconstitution of the hematolymphopoietic system in serial transfers. Such system provides a useful tool for practical application of iPSCs in the treatment of hematologic and immunologic diseases.

For further development of this method animal in which is grown the human cell graft, for example mouse, must have so modified genome that all its cells express and have on its surface human SIRPα. To prevent rejection after transplantation to the patient of the allogenic organ or tissue, grown from the pluripotent stem cells in vivo in the animal, these cells should express two molecules: CTLA4-Ig, which disrupts T cell costimulatory pathways and PD-L1, which activates T cell inhibitory pathway.

Differentiated Cell Types

Retinal Cells

In the near-future, clinical trials designed to demonstrate the safety of the use of iPSCs for cell therapy of the people with age-related macular degeneration, a disease causing blindness through retina damaging, will begin. There are several articles describing methods for producing retinal cells from iPSCs and how to use them for cell therapy. Reports of iPSC-derived retinal pigmented epithelium transplantation showed enhanced visual-guided behaviors of experimental animals for 6 weeks after transplantation. However, clinical trials have been successful: ten patients suffering from retinitis pigmentosa have had their eyesight restored—including a woman who had only 17 percent of her vision left.

Lung and Airway Epithelial Cells

Chronic lung diseases such as idiopathic pulmonary fibrosis and cystic fibrosis or chronic obstructive pulmonary disease and asthma are leading causes of morbidity and mortality worldwide with a considerable human, societal and financial burden. So there is an urgent need for effective cell therapy and lung tissue engineering. Several protocols have been developed for generation of the most cell types of the respiratory system, which may be useful for deriving patient-specific therapeutic cells.

Reproductive Cells

Some lines of iPSCs have the potentiality to differentiate into male germ cells and oocyte-like cells in an appropriate niche (by culturing in retinoic acid and porcine follicular fluid differentiation medium or seminiferous tubule transplantation). Moreover, iPSC transplantation make a contribution to repairing the testis of infertile mice, demonstrating the potentiality of gamete derivation from iPSCs in vivo and in vitro.

Induced Pluripotent Stem Cells

Direct Transdifferentiation

The risk of cancer and tumors creates the need to develop methods for safer cell lines suitable for clinical use. An alternative approach is so-called "direct reprogramming" - transdifferentiation of cells without passing through the pluripotent state. The basis for this approach was that 5-azacytidine - a DNA demethylation reagent - can cause the formation of myogenic, chondrogenic and adipogeni clones in the immortal cell line of mouse embryonic fibroblasts and that the activation of a single gene, later named MyoD1, is sufficient for such reprogramming. Compared with iPSC whose reprogramming requires at least two weeks, the formation of induced progenitor cells sometimes occurs within a few days and the efficiency of reprogramming is usually many times higher. This reprogramming does not always require cell division. The cells resulting from such reprogramming are more suitable for cell therapy because they do not form teratomas. For example, Chandrakanthan et al., & Pimanda describe the generation of tissue-regenerative multipotent stem cells (iMS cells) by treating mature bone and fat cells transiently with a growth factor (platelet-derived growth factor–AB (PDGF-AB)) and 5-Azacytidine. These authors claim that: "Unlike primary mesenchymal stem cells, which are used with little objective evidence in clinical practice to promote tissue repair, iMS cells contribute directly to in vivo tissue regeneration in a context-dependent manner without forming tumors" and so "has significant scope for application in tissue regeneration."

Single Transcription Factor Transdifferentiation

Originally only early embryonic cells could be coaxed into changing their identity. Mature cells are resistant to changing their identity once they've committed to a specific kind. However, brief expression of a single transcription factor, the ELT-7 GATA factor, can convert the identity of fully differentiated, specialized non-endodermal cells of the pharynx into fully differentiated intestinal cells in intact larvae and adult roundworm *Caenorhabditis elegans* with no requirement for a dedifferentiated intermediate.

Transdifferentiation with CRISPR-mediated Activator

The cell fate can be effectively manipulated by directly activating of specific endogenous gene expression with CRISPR-mediated activator. When dCas9 (that has been modified so that it no longer cuts DNA, but still can be guided to specific sequences and to bind to them) is combined with transcription activators, it can precisely manipulate endogenous gene expression. Using this method, Wei et al., enhanced the expression of endogenous Cdx2 and Gata6 genes by CRISPR-mediated activators, thus directly converted mouse embryonic stem cells into two extraembryonic lineages, i.e., typical trophoblast stem cells and extraembryonic endoderm cells. An analogous approach was used to induce activation of the endogenous Brn2, Ascl1, and Myt1l genes to convert mouse embryonic fibroblasts to induced neuronal cells. Thus, transcriptional activation and epigenetic remodeling of endogenous master transcription factors are sufficient for conversion between cell types. The rapid and sustained activation of endogenous genes in their native chromatin context by this approach may facilitate reprogramming with transient methods that avoid genomic integration and provides a new strategy for overcoming epigenetic barriers to cell fate specification.

Phased Process Modeling Regeneration

Another way of reprogramming is the simulation of the processes that occur during amphibian limb regeneration. In urodele amphibians, an early step in limb regeneration is skeletal muscle fiber dedifferentiation into a cellulate that proliferates into limb tissue. However, sequential small molecule treatment of the muscle fiber with myoseverin, reversine (the aurora B kinase inhibitor) and some other chemicals: BIO (glycogen synthase-3 kinase inhibitor), lysophosphatidic acid (pleiotropic activator of G-protein-coupled receptors), SB203580 (p38 MAP kinase inhibitor), or SQ22536 (adenylyl cyclase inhibitor) causes the formation of new muscle cell types as well as other cell types such as precursors to fat, bone and nervous system cells.

Antibody-based Transdifferentiation

The researchers discovered that GCSF-mimicking antibody can activate a growth-stimulating receptor on marrow cells in a way that induces marrow stem cells that normally develop into white blood cells to become neural progenitor cells. The technique enables researchers to search large libraries of antibodies and quickly select the ones with a desired biological effect.

Conditionally Reprogrammed Cells

Schlegel and Liu demonstrated that the combination of feeder cells and a Rho kinase inhibitor (Y-27632) induces normal and tumor epithelial cells from many tissues to proliferate indefinitely in vitro. This process occurs without the need for transduction of exogenous viral or cellular genes. These cells have been termed "Conditionally Reprogrammed Cells (CRC)". The induction of CRCs is rapid and results from reprogramming of the entire cell population. CRCs do not express high levels of proteins characteristic of iPSCs or embryonic stem cells (ESCs) (e.g., Sox2, Oct4, Nanog, or Klf4). This induction of CRCs is reversible and removal of Y-27632 and feeders allows the cells to differentiate normally. CRC technology can generate 2×10^6 cells in 5 to 6 days from needle biopsies and can generate cultures from cryopreserved tissue and from fewer than four viable cells. CRCs retain a normal karyotype and remain nontumorigenic. This technique also efficiently establishes cell cultures from human and rodent tumors.

The ability to rapidly generate many tumor cells from small biopsy specimens and frozen tissue provides significant opportunities for cell-based diagnostics and therapeutics (including chemosensitivity testing) and greatly expands the value of biobanking. Using CRC technology, researchers were able to identify an effective therapy for a patient with a rare type of lung tumor. Engleman's group describes a pharmacogenomic platform that facilitates rapid discovery of drug combinations that can overcome resistance using CRC system. In addition, the CRC method allows for the genetic manipulation of epithelial cells ex vivo and their subsequent evaluation in vivo in the same host. While initial studies revealed that co-culturing epithelial cells with Swiss 3T3 cells J2 was essential for CRC induction, with transwell culture plates, physical contact between feeders and epithelial cells is not required for inducing CRCs and more importantly that irradiation of the feeder cells is required for this induction. Consistent with the transwell experiments, conditioned medium induces and maintains CRCs, which is accompanied by a concomitant increase of cellular telomerase activity. The activity of the conditioned medium correlates directly with radiation-induced feeder cell apoptosis. Thus, conditional reprogramming of epithelial cells is mediated by a combination of Y-27632 and a soluble factor(s) released by apoptotic feeder cells.

Riegel et al. demonstrate that mouse ME cells isolated from normal mammary glands or from mouse mammary tumor virus (MMTV)-Neu–induced mammary tumors, can be cultured indefinitely as conditionally reprogrammed cells (CRCs). Cell surface progenitor-associated markers are rapidly induced in normal mouse ME-CRCs relative to ME cells. However, the expression of certain mammary progenitor subpopulations, such as CD49f+ ESA+ CD44+, drops significantly in later passages. Nevertheless, mouse ME-CRCs grown in a three-dimensional extracellular matrix gave rise to mammary acinar structures. ME-CRCs isolated from MMTV-Neu transgenic mouse mammary tumors express high levels of HER2/neu, as well as tumor-initiating cell markers, such as CD44+, CD49f+ and ESA+ (EpCam). These patterns of expression are sustained in later CRC passages. Early and late passage ME-CRCs from MMTV-Neu tumors that were implanted in the mammary fat pads of syngeneic or nude mice developed vascular tumors that metastasized within 6 weeks of transplantation. Importantly, the histopathology of these tumors was indistinguishable from that of the parental tumors that develop in the MMTV-Neu mice. Application of the CRC system to mouse mammary epithelial cells provides an attractive model system to study the genetics and phenotype of normal and transformed mouse epithelium in a defined culture environment and in vivo transplant studies.

A different approach to CRC is to inhibit CD47—a membrane protein that is the thrombospondin-1 receptor. Loss of CD47 permits sustained proliferation of primary murine endothelial cells, increases asymmetric division and enables these cells to spontaneously reprogram to form multipotent embryoid body-like clusters. CD47 knockdown acutely increases mRNA levels of c-Myc and other stem cell transcription factors in cells in vitro and in vivo. Thrombospondin-1 is a key environmental signal that inhibits stem cell self-renewal via CD47. Thus, CD47 antagonists enable cell self-renewal and reprogramming by overcoming negative regulation of c-Myc and other stem cell transcription factors. In vivo blockade of CD47 using an antisense morpholino increases survival of mice exposed to lethal total body irradiation due to increased proliferative capacity of bone marrow-derived cells and radioprotection of radiosensitive gastrointestinal tissues.

Lineage-specific Enhancers

Differentiated macrophages can self-renew in tissues and expand long-term in culture. Under certain conditions macrophages can divide without losing features they have acquired while specializing into immune cells - which is usually not possible with differentiated cells. The macrophages achieve this by activating a gene network similar to one found in embryonic stem cells. Single-cell analysis revealed that, *in vivo*, proliferating macrophages can derepress a macrophage-specific enhancer repertoire associated with a gene network controlling self-renewal. This happened when concentrations of two transcription factors named MafB and c-Maf were naturally low or were inhibited for a short time. Genetic manipulations that turned off MafB and c-Maf in the macrophages caused the cells to start a self-renewal program. The similar network also controls embryonic stem cell self-renewal but is associated with distinct embryonic stem cell-specific enhancers.

Hence macrophages isolated from MafB- and c-Maf-double deficient mice divide indefinitely; the self-renewal depends on c-Myc and Klf4.

Indirect Lineage Conversion

Indirect lineage conversion is a reprogramming methodology in which somatic cells transition through a plastic intermediate state of partially reprogrammed cells (pre-iP-SC), induced by brief exposure to reprogramming factors, followed by differentiation in a specially developed chemical environment (artificial niche).

This method could be both more efficient and safer, since it does not seem to produce tumors or other undesirable genetic changes and results in much greater yield than other methods. However, the safety of these cells remains questionable. Since lineage conversion from pre-iPSC relies on the use of iPSC reprogramming conditions, a fraction of the cells could acquire pluripotent properties if they do not stop the de-differentation process in vitro or due to further de-differentiation in vivo.

Outer Membrane Glycoprotein

A common feature of pluripotent stem cells is the specific nature of protein glycosylation of their outer membrane. That distinguishes them from most nonpluripotent cells, although not white blood cells. The glycans on the stem cell surface respond rapidly to alterations in cellular state and signaling and are therefore ideal for identifying even minor changes in cell populations. Many stem cell markers are based on cell surface glycan epitopes including the widely used markers SSEA-3, SSEA-4, Tra 1-60 and Tra 1-81. Suila Heli et al. speculate that in human stem cells extracellular O-GlcNAc and extracellular O-LacNAc, play a crucial role in the fine tuning of Notch signaling pathway - a highly conserved cell signaling system, that regulates cell fate specification, differentiation, left–right asymmetry, apoptosis, somitogenesis, angiogenesis and plays a key role in stem cell proliferation (reviewed by Perdigoto and Bardin and Jafar-Nejad et al.)

Changes in outer membrane protein glycosylation are markers of cell states connected in some way with pluripotency and differentiation. The glycosylation change is apparently not just the result of the initialization of gene expression, but perform as an important gene regulator involved in the acquisition and maintenance of the undifferentiated state.

For example, activation of glycoprotein ACA, linking glycosylphosphatidylinositol on the surface of the progenitor cells in human peripheral blood, induces increased expression of genes Wnt, Notch-1, BMI1 and HOXB4 through a signaling cascade PI3K/Akt/mTor/PTEN and promotes the formation of a self-renewing population of hematopoietic stem cells.

Furthermore, dedifferentiation of progenitor cells induced by ACA-dependent signaling pathway leads to ACA-induced pluripotent stem cells, capable of differentiating in vitro into cells of all three germ layers. The study of lectins' ability to maintain a culture of pluripotent human stem cells has led to the discovery of lectin Erythrina crista-galli (ECA), which can serve as a simple and highly effective matrix for the cultivation of human pluripotent stem cells.

Reprogramming Through a Physical Approach

Cell adhesion protein E-cadherin is indispensable for a robust pluripotent phenotype. During reprogramming for iPS cell generation, N-cadherin can replace function of E-cadherin. These functions of cadherins are not directly related to adhesion because sphere morphology helps maintaining the "stemness" of stem cells. Moreover, sphere formation, due to forced growth of cells on a low attachment surface, sometimes induces reprogramming. For example, neural progenitor cells can be generated from fibroblasts directly through a physical approach without introducing exogenous reprogramming factors.

Physical cues, in the form of parallel microgrooves on the surface of cell-adhesive substrates, can replace the effects of small-molecule epigenetic modifiers and significantly improve reprogramming efficiency. The mechanism relies on the mechanomodulation of the cells' epigenetic state. Specifically, "decreased histone deacetylase activity and upregulation of the expression of WD repeat domain 5 (WDR5)—a subunit of H3 methyltranferase—by microgrooved surfaces lead to increased histone H3 acetylation and methylation". Nanofibrous scaffolds with aligned fibre orientation produce effects similar to those produced by microgrooves, suggesting that changes in cell morphology may be responsible for modulation of the epigenetic state.

Substrate rigidity is an important biophysical cue influencing neural induction and subtype specification. For example, soft substrates promote neuroepithelial conversion while inhibiting neural crest differentiation of hESCs in a BMP4-dependent manner. Mechanistic studies revealed a multi-targeted mechanotransductive process involving mechanosensitive Smad phosphorylation and nucleocytoplasmic shuttling, regulated by rigidity-dependent Hippo/YAP activities and actomyosin cytoskeleton integrity and contractility.

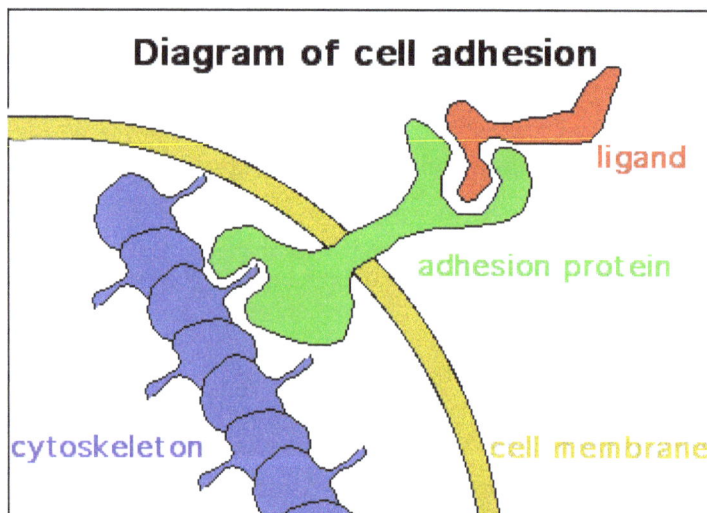

Role of cell adhesions in neural development. Image courtesy of Wikipedia user JWSchmidt under the GNU Free Documentation License

Mouse embryonic stem cells (mESCs) undergo self-renewal in the presence of the cytokine leukemia inhibitory factor (LIF). Following LIF withdrawal, mESCs differentiate, accompanied by an increase in cell–substratum adhesion and cell spreading. Restricted cell spreading in the absence of LIF by either culturing mESCs on chemically defined, weakly adhesive biosubstrates, or by manipulating the cytoskeleton allowed the cells to remain in an undifferentiated and pluripotent state. The effect of restricted cell spreading on mESC self-renewal is not mediated by increased intercellular adhesion, as inhibition of mESC adhesion using a function blocking anti E-cadherin antibody or siRNA does not promote differentiation. Possible mechanisms of stem cell fate predetermination by physical interactions with the extracellular matrix have been described.

A new method has been developed that turns cells into stem cells faster and more efficiently by 'squeezing' them using 3D microenvironment stiffness and density of the surrounding gel. The technique can be applied to a large number of cells to produce stem cells for medical purposes on an industrial scale.

Cells involved in the reprogramming process change morphologically as the process proceeds. This results in physical difference in adhesive forces among cells. Substantial differences in 'adhesive signature' between pluripotent stem cells, partially reprogrammed cells, differentiated progeny and somatic cells allowed to develop separation process for isolation of pluripotent stem cells in microfluidic devices, which is:

1. fast (separation takes less than 10 minutes);

2. efficient (separation results in a greater than 95 percent pure iPS cell culture);

3. innocuous (cell survival rate is greater than 80 percent and the resulting cells retain normal transcriptional profiles, differentiation potential and karyotype).

Stem cells possess mechanical memory (they remember past physical signals)—with the Hippo signaling pathway factors: Yes-associated protein (YAP) and transcriptional coactivator with PDZ-binding domain (TAZ) acting as an intracellular mechanical rheostat—that stores information from past physical environments and influences the cells' fate.

Neural Stem Cells

Stroke and many neurodegenerative disorders such as Parkinson's disease, Alzheimer's disease, amyotrophic lateral sclerosis need cell replacement therapy. The successful use of converted neural cells (cNs) in transplantations open a new avenue to treat such diseases. Nevertheless, induced neurons (iNs), directly converted from fibroblasts are terminally committed and exhibit very limited proliferative ability that may not provide enough autologous donor cells for transplantation. Self-renewing induced neural stem cells (iNSCs) provide additional advantages over iNs for both basic research and clinical applications.

For example, under specific growth conditions, mouse fibroblasts can be reprogrammed with a single factor, Sox2, to form iNSCs that self-renew in culture and after transplantation can survive and integrate without forming tumors in mouse brains. INSCs can be derived from adult human fibroblasts by non-viral techniques, thus offering a safe method for autologous transplantation or for the development of cell-based disease models.

Neural chemically induced progenitor cells (ciNPCs) can be generated from mouse tail-tip fibroblasts and human urinary somatic cells without introducing exogenous factors, but - by a chemical cocktail, namely VCR (V, VPA, an inhibitor of HDACs; C, CHIR99021, an inhibitor of GSK-3 kinases and R, RepSox, an inhibitor of TGF beta signaling pathways), under a physiological hypoxic condition. Alternative cocktails with inhibitors of histone deacetylation, glycogen synthase kinase and TGF-β pathways (where: sodium butyrate (NaB) or Trichostatin A (TSA) could replace VPA, Lithium chloride (LiCl) or lithium carbonate (Li2CO3) could substitute CHIR99021, or Repsox may be replaced with SB-431542 or Tranilast) show similar efficacies for ciNPC induction. Zhang, et al., also report highly efficient reprogramming of mouse fibroblasts into induced neural stem cell-like cells (ciNSLCs) using a cocktail of nine components.

Multiple methods of direct transformation of somatic cells into induced neural stem cells have been described.

Proof of principle experiments demonstrate that it is possible to convert transplanted human fibroblasts and human astrocytes directly in the brain that are engineered to express inducible forms of neural reprogramming genes, into neurons, when reprogramming genes (Ascl1, Brn2a and Myt1l) are activated after transplantation using a drug.

Astrocytes—the most common neuroglial brain cells, which contribute to scar formation in response to injury—can be directly reprogrammed in vivo to become functional neurons that formed networks in mice without the need of cell transplantation. The researchers followed the mice for nearly a year to look for signs of tumor formation and reported finding none. The same researchers have turned scar-forming astrocytes into progenitor cells called neuroblasts that regenerated into neurons in the injured adult spinal cord.

Oligodendrocyte Precursor Cells

Without myelin to insulate neurons, nerve signals quickly lose power. Diseases that attack myelin, such as multiple sclerosis, result in nerve signals that cannot propagate to nerve endings and as a consequence lead to cognitive, motor and sensory problems. Transplantation of oligodendrocyte precursor cells (OPCs), which can successfully create myelin sheaths around nerve cells, is a promising potential therapeutic response. Direct lineage conversion of mouse and rat fibroblasts into oligodendroglial cells provides a potential source of OPCs. Conversion by forced expression of both eight or of the three transcription factors Sox10, Olig2 and Zfp536, may provide such cells.

Cardiomyocytes

Cell-based in vivo therapies may provide a transformative approach to augment vascular and muscle growth and to prevent non-contractile scar formation by delivering transcription factors or microRNAs to the heart. Cardiac fibroblasts, which represent 50% of the cells in the mammalian heart, can be reprogrammed into cardiomyocyte-like cells in vivo by local delivery of cardiac core transcription factors (GATA4, MEF2C, TBX5 and for improved reprogramming plus ESRRG, MESP1, Myocardin and ZFPM2) after coronary ligation. These results implicated therapies that can directly remuscularize the heart without cell transplantation. However, the efficiency of such reprogramming turned out to be very low and the phenotype of received cardiomyocyte-like cells does not resemble those of a mature normal cardiomyocyte. Furthermore, transplantation of cardiac transcription factors into injured murine hearts resulted in poor cell survival and minimal expression of cardiac genes.

Meanwhile, advances in the methods of obtaining cardiac myocytes in vitro occurred. Efficient cardiac differentiation of human iPS cells gave rise to progenitors that were retained within infarcted rat hearts and reduced remodeling of the heart after ischemic damage.

The team of scientists, who were led by Sheng Ding, used a cocktail of nine chemicals (9C) for transdifferentiation of human skin cells into beating heart cells. With this method, more than 97% of the cells began beating, a characteristic of fully developed, healthy heart cells. The chemically induced cardiomyocyte-like cells (ciCMs) uniformly contracted and resembled human cardiomyocytes in their transcriptome, epigenetic, and electrophysiological properties. When transplanted into infarcted mouse hearts, 9C-treated fibroblasts were efficiently converted to ciCMs and developed into healthy-looking heart muscle cells within the organ. This chemical reprogramming approach, after further optimization, may offer an easy way to provide the cues that induce heart muscle to regenerate locally.

In another study, ischemic cardiomyopathy in the murine infarction model was targeted by iPS cell transplantation. It synchronized failing ventricles, offering a regenerative strategy to achieve resynchronization and protection from decompensation by dint of improved left ventricular conduction and contractility, reduced scarring and reversal of structural remodelling. One protocol generated populations of up to 98% cardiomyocytes from hPSCs simply by modulating the canonical Wnt signaling pathway at defined time points in during differentiation, using readily accessible small molecule compounds.

Discovery of the mechanisms controlling the formation of cardiomyocytes led to the development of the drug ITD-1, which effectively clears the cell surface from TGF-β receptor type II and selectively inhibits intracellular TGF-β signaling. It thus selectively enhances the differentiation of uncommitted mesoderm to cardiomyocytes, but not to vascular smooth muscle and endothelial cells.

One project seeded decellularized mouse hearts with human iPSC-derived multipotential cardiovascular progenitor cells. The introduced cells migrated, proliferated and differentiated in situ into cardiomyocytes, smooth muscle cells and endothelial cells to reconstruct the hearts. In addition, the heart's extracellular matrix (the substrate of heart scaffold) signalled the human cells into becoming the specialised cells needed for proper heart function. After 20 days of perfusion with growth factors, the engineered heart tissues started to beat again and were responsive to drugs.

Rejuvenation of the Muscle Stem Cell

The elderly often suffer from progressive muscle weakness and regenerative failure owing in part to elevated activity of the p38α and p38β mitogen-activated kinase pathway in senescent skeletal muscle stem cells. Subjecting such stem cells to transient inhibition of p38α and p38β in conjunction with culture on soft hydrogel substrates rapidly expands and rejuvenates them that result in the return of their strength.

In geriatric mice, resting satellite cells lose reversible quiescence by switching to an irreversible pre-senescence state, caused by derepression of p16INK4a (also called Cdkn2a). On injury, these cells fail to activate and expand, even in a youthful environment. p16INK4a silencing in geriatric satellite cells restores quiescence and muscle regenerative functions.

Myogenic progenitors for potential use in disease modeling or cell-based therapies targeting skeletal muscle could also be generated directly from induced pluripotent stem cells using free-floating spherical culture (EZ spheres) in a culture medium supplemented with high concentrations (100 ng/ml) of fibroblast growth factor-2 (FGF-2) and epidermal growth factor.

Hepatocytes

An intestinal crypt - an accessible and abundant source of intestinal epithelial cells for conversion into β-like cells.

Unlike current protocols for deriving hepatocytes from human fibroblasts, Saiyong Zhu et al., (2014) did not generate iPSCs but, using small molecules, cut short reprogramming to pluripotency to generate an induced multipotent progenitor cell (iMPC) state from which endoderm progenitor cells and subsequently hepatocytes (iMPC-Heps) were efficiently differentiated. After transplantation into an immune-deficient mouse model of human liver failure, iMPC-Heps proliferated extensively and acquired levels of hepatocyte function similar to those of human primary adult hepatocytes. iMPC-Heps did not form tumours, most probably because they never entered a pluripotent state.

These results establish the feasibility of significant liver repopulation of mice with human hepatocytes generated in vitro, which removes a long-standing roadblock on the path to autologous liver cell therapy.

Insulin-producing Cells

Complications of Diabetes mellitus such as cardiovascular diseases, retinopathy, neuropathy, nephropathy and peripheral circulatory diseases depend on sugar dysregulation due to lack of insulin from pancreatic beta cells and can be lethal if they are not treated. One of the promising approaches to understand and cure diabetes is to use pluripotent stem cells (PSCs), including embryonic stem cells (ESCs) and induced PCSs (iPSCs). Unfortunately, human PSC-derived insulin-expressing cells resemble human fetal β cells rather than adult β cells. In contrast to adult β cells, fetal β cells seem functionally immature, as indicated by increased basal glucose secretion and lack of glucose stimulation and confirmed by RNA-seq of whose transcripts.

An alternative strategy is the conversion of fibroblasts towards distinct endodermal progenitor cell populations and, using cocktails of signalling factors, successful differentiation of these endodermal progenitor cells into functional beta-like cells both in vitro and in vivo.

Overexpression of the three transcription factors, PDX1 (required for pancreatic bud outgrowth and beta-cell maturation), NGN3 (required for endocrine precursor cell formation) and MAFA (for beta-cell maturation) combination (called PNM) can lead to the transformation of some cell types into a beta cell-like state. An accessible and abundant source of functional insulin-producing cells is intestine. PMN expression in human intestinal "organoids" stimulates the conversion of intestinal epithelial cells into β-like cells possibly acceptable for transplantation.

Nephron Progenitors

Adult proximal tubule cells were directly transcriptionally reprogrammed to nephron progenitors of the embryonic kidney, using a pool of six genes of instructive transcription factors (SIX1, SIX2, OSR1, Eyes absent homolog 1(EYA1), Homeobox A11 (HOXA11) and Snail homolog 2 (SNAI2)) that activated genes consistent with a cap mesenchyme/nephron progenitor phenotype in the adult proximal tubule cell line. The generation of such cells may lead to cellular therapies for adult renal disease. Embryonic kidney organoids placed into adult rat kidneys can undergo onward development and vascular development.

Blood Vessel Cells

As blood vessels age, they often become abnormal in structure and function, thereby contributing to numerous age-associated diseases including myocardial infarction, ischemic stroke and atherosclerosis of arteries supplying the heart, brain and lower extremities. So, an important goal is to stimulate vascular growth for the collateral circulation to prevent the exacerbation of these diseases. Induced Vascular Progenitor Cells (iVPCs) are useful for cell-based therapy designed to stimulate coronary collateral growth. They were generated by partially reprogramming endothelial cells. The vascular commitment of iVPCs is related to the epigenetic memory of endothelial cells, which engenders them as cellular components of growing blood vessels. That is why, when iVPCs were implanted into myocardium, they engrafted in blood vessels and increased coronary collateral flow better than iPSCs, mesenchymal stem cells, or native endothelial cells.

Ex vivo genetic modification can be an effective strategy to enhance stem cell function. For example, cellular therapy employing genetic modification with Pim-1 kinase (a downstream effector of Akt, which positively regulates neovasculogenesis) of bone marrow–derived cells or human cardiac progenitor cells, isolated from failing myocardium results in durability of repair, together with the improvement of functional parameters of myocardial hemodynamic performance.

Stem cells extracted from fat tissue after liposuction may be coaxed into becoming progenitor smooth muscle cells (iPVSMCs) found in arteries and veins.

The 2D culture system of human iPS cells in conjunction with triple marker selection (CD34 (a surface glycophosphoprotein expressed on developmentally early embryonic fibroblasts), NP1 (receptor - neuropilin 1) and KDR (kinase insert domain-containing receptor)) for the isolation of vasculogenic precursor cells from human iPSC, generated endothelial cells that, after transplantation, formed stable, functional mouse blood vessels in vivo, lasting for 280 days.

To treat infarction, it is important to prevent the formation of fibrotic scar tissue. This can be achieved in vivo by transient application of paracrine factors that redirect native heart progenitor stem cell contributions from scar tissue to cardiovascular tissue. For example, in a mouse myocardial infarction model, a single intramyocardial injection of human vascular endothelial growth factor A mRNA (VEGF-A modRNA), modified to escape the body's normal defense system, results in long-term improvement of heart function due to mobilization and redirection of epicardial progenitor cells toward cardiovascular cell types.

Blood Stem Cells

Red Blood Cells

RBC transfusion is necessary for many patients. However, to date the supply of RBCs remains labile. In addition, transfusion risks infectious disease transmission. A large supply of safe RBCs generated in vitro would help to address this issue. Ex vivo erythroid cell generation may provide alternative transfusion products to meet present and future clinical requirements. Red blood cells (RBC)s generated in vitro from mobilized CD34 positive cells have normal survival when transfused into an autologous recipient. RBC produced in vitro contained exclusively fetal hemoglobin (HbF), which rescues the functionality of these RBCs. In vivo the switch of fetal to adult hemoglobin was observed after infusion of nucleated erythroid precursors derived from iPSCs. Although RBCs do not have nuclei and therefore can not form a tumor, their immediate erythroblasts precursors have nuclei. The terminal maturation of erythroblasts into functional RBCs requires a complex remodeling process that ends with extrusion of the nucleus and the formation of an enucleated RBC. Cell reprogramming often disrupts enucleation. Transfusion of in vitro-generated RBCs or erythroblasts does not sufficiently protect against tumor formation.

The aryl hydrocarbon receptor (AhR) pathway (which has been shown to be involved in the promotion of cancer cell development) plays an important role in normal blood cell development. AhR activation in human hematopoietic progenitor cells (HPs) drives an unprecedented expansion of HPs, megakaryocyte- and erythroid-lineage cells. The SH2B3 gene encodes a negative regulator of cytokine signaling and naturally occurring loss-of-function variants in this gene increase RBC counts in vivo. Targeted sup-pression of SH2B3 in primary human hematopoietic stem and progenitor cells enhanced the maturation and overall yield of in-vitro-derived RBCs. Moreover, inactivation of SH2B3 by CRISPR/Cas9 genome editing in human pluripotent stem cells allowed enhanced erythroid cell expansion with preserved differentiation.

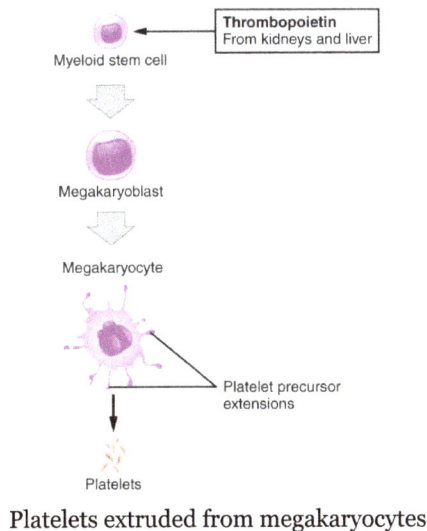

Platelets extruded from megakaryocytes

Platelets

Platelets help prevent hemorrhage in thrombocytopenic patients and patients with thrombocythemia. A significant problem for multitransfused patients is refractoriness to platelet transfusions. Thus, the ability to generate platelet products ex vivo and platelet products lacking HLA antigens in serum-free media would have clinical value. An RNA interference-based mechanism used a lentiviral vector to express short-hairpin RNAi targeting β2-microglobulin transcripts in CD34-positive cells. Generated platelets demonstrated an 85% reduction in class I HLA antigens. These platelets appeared to have normal function in vitro

One clinically-applicable strategy for the derivation of functional platelets from human iPSC involves the establishment of stable immortalized megakaryocyte progenitor cell lines (imMKCLs) through doxycycline-dependent overexpression of BMI1 and BCL-XL. The resulting imMKCLs can be expanded in culture over extended periods (4–5 months), even after cryopreservation. Halting the overexpression (by the removal of doxycycline from the medium) of c-MYC, BMI1 and BCL-XL in growing imMKCLs led to the production of CD42b+ platelets with functionality comparable to that of native platelets on the basis of a range of assays in vitro and in vivo. Thomas et al., describe a forward programming strategy relying on the concurrent exogenous expression of 3 transcription factors: GATA1, FLI1 and TAL1. The forward programmed megakaryocytes proliferate and differentiate in culture for several months with megakaryocyte purity over 90% reaching up to 2×10^5 mature megakaryocytes per input hPSC. Functional platelets are generated throughout the culture allowing the prospective collection of several transfusion units from as few as one million starting hPSCs.

Immune Cells

A specialised type of white blood cell, known as cytotoxic T lymphocytes (CTLs), are produced by the immune system and are able to recognise specific markers on the surface of various infectious or tumour cells, causing them to launch an attack to kill the harmful cells. Thence, immunotherapy with functional antigen-specific T cells has potential as a therapeutic strategy for combating many cancers and viral infections. However, cell sources are limited, because they are produced in small numbers naturally and have a short lifespan.

A potentially efficient approach for generating antigen-specific CTLs is to revert mature immune T cells into iPSCs, which possess indefinite proliferative capacity in vitro and after their multiplication to coax them to redifferentiate back into T cells.

Another method combines iPSC and chimeric antigen receptor (CAR) technologies to generate human T cells targeted to CD19, an antigen expressed by malignant B cells, in tissue culture. This approach of generating therapeutic human T cells may be useful for cancer immunotherapy and other medical applications.

Invariant natural killer T (iNKT) cells have great clinical potential as adjuvants for cancer immunotherapy. iNKT cells act as innate T lymphocytes and serve as a bridge between the innate and acquired immune systems. They augment anti-tumor responses by producing interferon-gamma (IFN-γ). The approach of collection, reprogramming/dedifferentiation, re-differentiation and injection has been proposed for related tumor treatment.

Dendritic cells (DC) are specialized to control T-cell responses. DC with appropriate genetic modifications may survive long enough to stimulate antigen-specific CTL and after that be completely eliminated. DC-like antigen-presenting cells obtained from human induced pluripotent stem cells can serve as a source for vaccination therapy.

CCAAT/enhancer binding protein-α (C/EBPα) induces transdifferentiation of B cells into macrophages at high efficiencies and enhances reprogramming into iPS cells when co-expressed with transcription factors Oct4, Sox2, Klf4 and Myc. with a 100-fold increase in iPS cell reprogramming efficiency, involving 95% of the population. Furthermore, C/EBPa can convert selected human B cell lymphoma and leukemia cell lines into macrophage-like cells at high efficiencies, impairing the cells' tumor-forming capacity.

Thymic Epithelial Cells Rejuvenation

The thymus is the first organ to deteriorate as people age. This shrinking is one of the main reasons the immune system becomes less effective with age. Diminished expression of the thymic epithelial cell transcription factor FOXN1 has been implicated as a component of the mechanism regulating age-related involution.

Clare Blackburn and colleagues show that established age-related thymic involution can be reversed by forced upregulation of just one transcription factor - FOXN1 in the thymic epithelial cells in order to promote rejuvenation, proliferation and differentiation of these cells into fully functional thymic epithelium. This rejuvenation and increased proliferation was accompanied by upregulation of genes that promote cell cycle progression (cyclin D1, ΔNp63, FgfR2IIIb) and that are required in the thymic epithelial cells to promote specific aspects of T cell development (Dll4, Kitl, Ccl25, Cxcl12, Cd40, Cd80, Ctsl, Pax1).

Mesenchymal Stem Cells

Induction

mesenchymal stem/stromal cells (MSCs) are under investigation for applications in cardiac, renal, neural, joint and bone repair, as well as in inflammatory conditions and hemopoietic cotransplantation. This is because of their immunosuppressive properties and ability to differentiate into a wide range of mesenchymal-lineage tissues. MSCs are typically harvested from adult bone marrow or fat, but these require painful invasive procedures and are low-frequency sources, making up only 0.001%– 0.01% of bone marrow cells and 0.05% in liposuction aspirates. Of concern for autologous use, in particular in the elderly most in need of tissue repair, MSCs decline in quantity and quality with age.

IPSCs could be obtained by the cells rejuvenation of even centenarians. Because iPSCs can be harvested free of ethical constraints and culture can be expanded indefinitely, they are an advantageous source of MSCs. IPSC treatment with SB-431542 leads to rapid and uniform MSC generation from human iPSCs. (SB-431542 is an inhibitor of activin/TGF- pathways by blocking phosphorylation of ALK4, ALK5 and ALK7 receptors.) These iPS-MSCs may lack teratoma-forming ability, display a normal stable karyotype in culture and exhibit growth and differentiation characteristics that closely resemble those of primary MSCs. It has potential for in vitro scale-up, enabling MSC-based therapies. MSC derived from iPSC have the capacity to aid periodontal regeneration and are a promising source of readily accessible stem cells for use in the clinical treatment of periodontitis.

Besides cell therapy in vivo, the culture of human mesenchymal stem cells can be used in vitro for mass-production of exosomes, which are ideal vehicles for drug delivery.

Dedifferentiated Adipocytes

Adipose tissue, because of its abundance and relatively less invasive harvest methods, represents a source of mesenchymal stem cells (MSCs). Unfortunately, liposuction aspirates are only 0.05% MSCs. However, a large amount of mature adipocytes, which in general have lost their proliferative abilities and therefore are typically discarded, can be easily isolated from the adipose cell suspension and dedifferentiated into lipid-free fibroblast-like cells, named dedifferentiated fat (DFAT) cells. DFAT cells re-establish active proliferation ability and express multipotent capacities. Compared with adult stem cells, DFAT cells show unique advantages in abundance, isolation and homogeneity. Under proper induction culture in vitro or proper environment in vivo, DFAT cells could demonstrate adipogenic, osteogenic, chondrogenic and myogenic potentials. They could also exhibit perivascular characteristics and elicit neovascularization.

Chondrogenic Cells

Cartilage is the connective tissue responsible for frictionless joint movement. Its degeneration ultimately results in complete loss of joint function in the late stages of osteoarthritis. As an avascular and hypocellular tissue, cartilage has a limited capacity for self-repair. Chondrocytes are the only cell type in cartilage, in which they are surrounded by the extracellular matrix that they secrete and assemble.

One method of producing cartilage is to induce it from iPS cells. Alternatively, it is possible to convert fibroblasts directly into induced chondrogenic cells (iChon) without an intermediate iPS cell stage, by inserting three reprogramming factors (c-MYC, KLF4 and SOX9). Human iChon cells expressed marker genes for chondrocytes (type II collagen) but not fibroblasts.

Implanted into defects created in the articular cartilage of rats, human iChon cells survived to form cartilaginous tissue for at least four weeks, with no tumors. The method makes use of c-MYC, which is thought to have a major role in tumorigenesis and employs a retrovirus to introduce the reprogramming factors, excluding it from unmodified use in human therapy.

Sources of Cells for Reprogramming

The most frequently used sources for reprogramming are blood cells and fibroblasts, obtained by biopsy of the skin, but taking cells from urine is less invasive. The latter method does not require a biopsy or blood sampling. As of 2013, urine-derived stem cells had been differentiated into endothelial, osteogenic, chondrogenic, adipogenic, skeletal myogenic and neurogenic lineages, without forming teratomas. Therefore, their epigenetic memory is suited to reprogramming into iPS cells. However, few cells appear in urine, only low conversion efficiencies had been achieved and the risk of bacterial contamination is relatively high.

Another promising source of cells for reprogramming are mesenchymal stem cells derived from human hair follicles.

The origin of somatic cells used for reprogramming may influence the efficiency of reprogramming, the functional properties of the resulting induced stem cells and the ability to form tumors.

IPSCs retain an epigenetic memory of their tissue of origin, which impacts their differentiation potential. This epigenetic memory does not necessarily manifest itself at the pluripotency stage – iPSCs derived from different tissues exhibit proper morphology, express pluripotency markers and are able to differentiate into the three embryonic layers in vitro and in vivo. However, this epigenetic memory may manifest during re-differentiation into specific cell types that require the specific loci bearing residual epigenetic marks.

Progenitor Cell

A progenitor cell is a biological cell that, like a stem cell, has a tendency to differentiate into a specific type of cell, but is already more specific than a stem cell and is pushed to differentiate into its "target" cell. The most important difference between stem cells and progenitor cells is that stem cells can replicate indefinitely, whereas progenitor cells can divide only a limited number of times. Controversy about the exact definition remains and the concept is still evolving.

The terms "progenitor cell" and "stem cell" are sometimes equated.

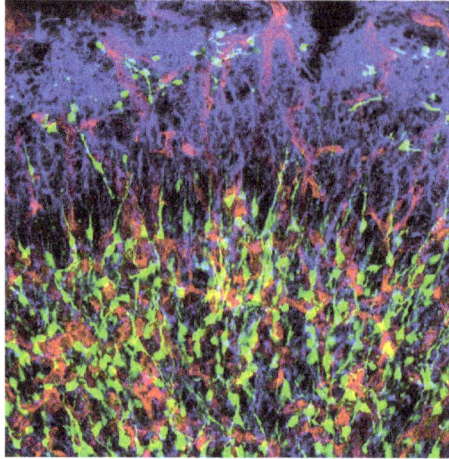

Neural progenitors (green) in olfactory bulb with astrocytes (blue).

Properties

Most progenitors are described as oligopotent. In this point of view, they may be compared to adult stem cells. But progenitors are said to be in a further stage of cell differentiation. They are in the "center" between stem cells and fully differentiated cells. The kind of potency they have depends on the type of their "parent" stem cell and also on their niche. Some progenitor cells were found during research, and were isolated. After their marker was found, it was proven that these progenitor could move through the body and migrate towards the tissue where they are needed. Many properties are shared by adult stem cells and progenitor cells.

Progenitor cells are found in adult organisms and they act as a repair system for the body. They replenish special cells, but also maintain the blood, skin and intestinal tissues. They can also be found in developing embryonic pancreatic tissue.

	Stem Cell	Progenitor Cell
	Self-renewal in vitro	
	Unlimited	
	Limited	
Potentiality	Multipotent	Unipotent, sometimes oligopotent
Maintenance of self-renewal	Yes	No
Population	Reaches maximum number of cells before differentiating	Does not reach maximum population

Function

The majority of progenitor cells lie dormant or possess little activity in the tissue in which they reside. They exhibit slow growth and their main role is to replace cells lost by normal attrition. In case of tissue injury, damaged or dead cells, progenitor cells can be activated. Growth factors or cytokines are two substances that trigger the progenitors to mobilize toward the damaged tissue. At the same time, they start to differentiate into the target cells. Not all progenitors are mobile and are situated near the tissue of their target differentiation. When the cytokines, growth factors and other cell division enhancing stimulators take on the progenitors, a higher rate of cell division is introduced. It leads to the recovery of the tissue.

Examples

The characterization or the defining principle of progenitor cells, in order to separate them from others, is based on the different cell markers rather than their morphological appearance.

- Satellite cells found in muscles. They play a major role in muscle cell differentiation and injury recoveries.

- Intermediate progenitor cells formed in the subventricular zone. Some of these transit amplifying neural progenitors migrate via rostral migratory stream to the olfactory bulb and differentiate further into specific types of neural cells.

- Bone marrow stromal cells, basal cell of epidermis have 10% of progenitor cell, although they are often classed as stem cells due to their high plasticity and potential for unlimited capacity for self-renewal.

- Periosteum contains progenitor cells that develop into osteoblasts and chondroblasts.

- Pancreatic progenitor cells are among the most-studied progenitors. They are used in research to develop a cure against diabetes type-1.

- Angioblasts or endothelial progenitor cells (EPC). These are very important for research on fracture and wounds healing.

- Blast cells are involved in generation of B- and T-lymphocytes, which participate in immune responses.

Development of the Human Cerebral Cortices

Before embryonic day 40 (E40), progenitor cells generate other progenitor cells; after that period, progenitor cells produce only dissimilar mesenchymal stem cell daughters. The cells from a single progenitor cell form a proliferative unit that creates one cortical column; these columns contain a variety of neurons with different shapes.

Hematopoietic Stem Cell

Hematopoietic stem cells (HSCs) or hemocytoblasts are the stem cells that give rise to all the other blood cells through the process of haematopoiesis. They are derived from mesoderm and located in the red bone marrow, which is contained in the core of most bones.

They give rise to both the myeloid and lymphoid lineages of blood cells. (Myeloid cells include monocytes, macrophages, neutrophils, basophils, eosinophils, erythrocytes, dendritic cells, and megakaryocytes or platelets. Lymphoid cells include T cells, B cells, and natural killer cells.) The definition of hematopoietic stem cells has evolved since HSCs were first discovered in 1961. The hematopoietic tissue contains cells with long-term and short-term regeneration capacities and committed multipotent, oligopotent, and unipotent progenitors. HSCs constitute 1:10.000 of cells in myeloid tissue.

HSCs are a heterogeneous population. The third category consists of the balanced (Bala) HSC, whose L/M ratio is between 3 and 10. Only the myeloid-biased and -balanced HSCs have durable self-renewal properties. In addition, serial transplantation experiments have shown that each subtype preferentially re-creates its blood cell type distribution, suggesting an inherited epigenetic program for each subtype.

HSC studies through much of the past half century have led to a much deeper understanding. More recent advances have resulted in the use of HSC transplants in the treatment of cancers and other immune system disorders.

Sources

HSCs are found in the bone marrow of adults, specially in the pelvis, femur, and sternum. They are also found in umbilical cord blood and, in small numbers, in peripheral blood.

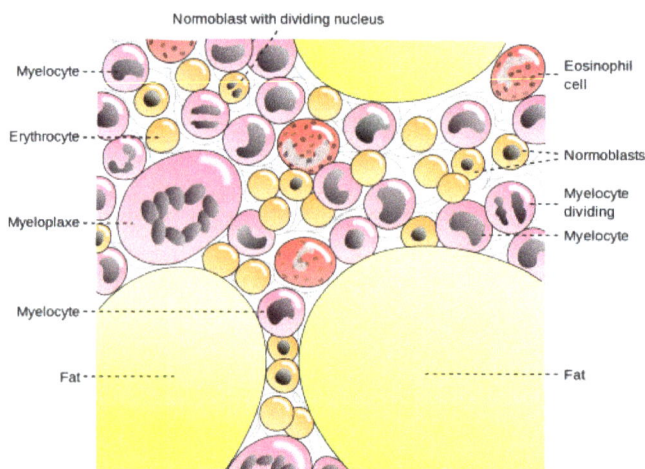

Sketch of bone marrow and its cells

Stem and progenitor cells can be taken from the pelvis, at the iliac crest, using a needle and syringe. The cells can be removed as liquid (to perform a smear to look at the cell morphology) or they can be removed via a core biopsy (to maintain the architecture or relationship of the cells to each other and to the bone).In order to harvest stem cells from the circulating peripheral blood, blood donors are injected with a cytokine, such as granulocyte-colony stimulating factor (G-CSF), that induces cells to leave the bone marrow and circulate in the blood vessels.In mammalian embryology, the first definitive HSCs are detected in the AGM (aorta-gonad-mesonephros), and then massively expanded in the fetal liver prior to colonising the bone marrow before birth.

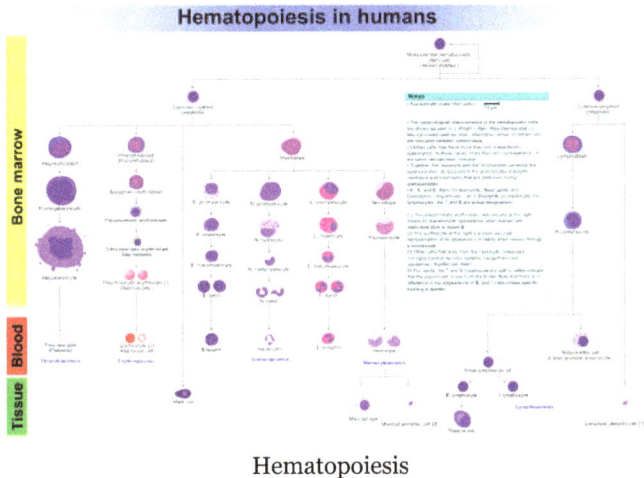

Hematopoiesis

Functional Characteristics

Multipotency and Self-renewal

HSCs can replenish all blood cell types (i.e., are multipotent) and self-renew. A small number of HSCs can expand to generate a very large number of daughter HSCs. This phenomenon is used in bone marrow transplantation, when a small number of HSCs reconstitute the hematopoietic system. This process indicates that, subsequent to bone marrow transplantation, symmetrical cell divisions into two daughter HSCs must occur.

Stem cell self-renewal is thought to occur in the stem cell niche in the bone marrow, and it is reasonable to assume that key signals present in this niche will be important in self-renewal. There is much interest in the environmental and molecular requirements for HSC self-renewal, as understanding the ability of HSC to replenish themselves will eventually allow the generation of expanded populations of HSC *in vitro* that can be used therapeutically.

Stem Cell Heterogeneity

It was originally believed that all HSCs were alike in their self-renewal and differentiation abilities. This view was first challenged by the 2002 discovery by the Muller-Sieburg group in San Diego, who illustrated that different stem cells can show distinct re-population patterns that are epigenetically predetermined intrinsic properties of clonal Thy-1^{lo} Sca-1$^+$ lin$^-$ c-kit$^+$ HSC. The results of these clonal studies led to the notion of lineage bias. Using the ratio $\rho = L / M$ of lymphoid (L) to myeloid (M) cells in blood as a quantitative marker, the stem cell compartment can be split into three categories of HSC. Balanced (Bala) HSCs repopulate peripheral white blood cells in the same ratio of myeloid to lymphoid cells as seen in unmanipulated mice (on average about 15% myeloid and 85% lymphoid cells, or $3 \leq \rho \leq 10$). Myeloid-biased (My-bi) HSCs give rise to very few lymphocytes resulting in ratios $0 < \rho < 3$, while lymphoid-biased (Ly-bi) HSCs generate very few myeloid cells, which results in lymphoid-to-myeloid ratios of $\rho > 10$. All three types are normal types of HSC, and they do not represent stages of differentiation. Rather, these are three classes of HSC, each with an epigenetically fixed differentiation program. These studies also showed that lineage bias is not stochasti-cally regulated or dependent on differences in environmental influence. My-bi HSC self-renew longer than balanced or Ly-bi HSC. The myeloid bias results from reduced responsiveness to the lymphopoetin interleukin 7 (IL-7).

Subsequently, other groups confirmed and highlighted the original findings. For exam-ple, the Eaves group confirmed in 2007 that repopulation kinetics, long-term self-re-newal capacity, and My-bi and Ly-bi are stably inherited intrinsic HSC properties. In 2010, the Goodell group provided additional insights about the molecular basis of lineage bias in side population (SP) SCA-1$^+$ lin$^-$ c-kit$^+$ HSC. As previously shown for IL-7 signaling, it was found that a member of the transforming growth factor family (TGF-beta) induces and inhibits the proliferation of My-bi and Ly-bi HSC, respectively.

Behavior in Culture

A *cobblestone area-forming cell (CAFC)* assay is a cell culture-based empirical assay. When plated onto a confluent culture of stromal feeder layer, a fraction of HSCs creep between the gaps (even though the stromal cells are touching each other) and even-tually settle between the stromal cells and the substratum (here the dish surface) or trapped in the cellular processes between the stromal cells. Emperipolesis is the *in vivo* phenomenon in which one cell is completely engulfed into another (e.g. thymocytes into thymic nurse cells); on the other hand, when *in vitro*, lymphoid lineage cells creep beneath nurse-like cells, the process is called pseudoemperipolesis. This similar phe-nomenon is more commonly known in the HSC field by the cell culture terminology *cobble stone area-forming cells (CAFC)*, which means areas or clusters of cells look dull cobblestone-like under phase contrast microscopy, compared to the other HSCs, which are refractile. This happens because the cells that are floating loosely on top of the stromal cells are spherical and thus refractile. However, the cells that creep beneath the stromal cells are flattened and, thus, not refractile. The mechanism of pseudoem-peripolesis is only recently coming to light. It may be mediated by interaction through CXCR4 (CD184) the receptor for CXC Chemokines (e.g., SDF1) and $\alpha 4 \beta 1$ integrins.

Mobility

HSCs have a higher potential than other immature blood cells to pass the bone marrow barrier, and, thus, may travel in the blood from the bone marrow in one bone to another bone. If they settle in the thymus, they may develop into T cells. In the case of fetuses and other extramedullary hematopoiesis, HSCs may also settle in the liver or spleen and develop.

This enables HSCs to be harvested directly from the blood.

Physical Characteristics

With regard to morphology, hematopoietic stem cells resemble lymphocytes. They are non-adherent, and rounded, with a rounded nucleus and low cytoplasm-to-nucleus ratio. Since primitive hematopoietic stem cells (PHSCs) cannot be isolated as a pure population, it is not possible to identify them in a microscope. The above description is based on the morphological characteristics of a heterogeneous population, of which PHSCs are a component.

Markers

In reference to phenotype, hematopoeitic stem cells are identified by their small size, lack of lineage (lin) markers, low staining (side population) with vital dyes such as rhodamine 123 (rhodamine^{DULL}, also called rho^{lo}) or Hoechst 33342, and presence of various antigenic markers on their surface.

Cluster of Differentiation and Other Markers

The classical marker of human HSC is CD34 first described independently by Civin et al. and Tindle et al. It is used to isolate HSC for reconstitution of patients who are haematologically incompetent as a result of chemotherapy or disease.

Many markers belong to the cluster of differentiation series, like: CD34, CD38, CD90, CD133, CD105, CD45, and also c-kit, - the receptor for stem cell factor. The hematopoietic stem cells are negative for the markers that are used for detection of lineage commitment, and are, thus, called Lin-; and, during their purification by FACS, a mixture of up to 14 different mature blood-lineage-marker antibodies are used to deplete the lin+ cells or late multipotent progenitors (MPP)s: e.g., CD13 & CD33 for myeloid, CD71 for erythroid, CD19 for B cells, CD61 for megakaryocytic, etc. for humans; and, B220 (murine CD45) for B cells, Mac-1 (CD11b/CD18) for monocytes, Gr-1 for Granulocytes, Ter119 for erythroid cells, Il7Ra, CD3, CD4, CD5, CD8 for T cells, etc. (for mice)

There are many differences between the human and mice hematopoietic cell markers for the commonly accepted type of hematopoietic stem cells.

- Mouse HSC: CD34$^{lo/-}$, SCA-1$^+$, Thy1.1$^{+/lo}$, CD38$^+$, C-kit$^+$, lin$^-$

- Human HSC: CD34$^+$, CD59$^+$, Thy1/CD90$^+$, CD38$^{lo/-}$, C-kit/CD117$^+$, lin$^-$

However, not all stem cells are covered by these combinations that, nonetheless, have become popular. In fact, even in humans, there are hematopoietic stem cells that are CD34$^-$/CD38$^-$. Also some later studies suggested that earliest stem cells may lack c-kit on the cell surface. For human HSCs use of CD133 was one step ahead as both CD34$^+$ and CD34$^-$ HSCs were CD133$^+$.

Traditional purification method used to yield a reasonable purity level of mouse hematopoietic stem cells, in general, requires a large(~10-12) battery of markers, most of which were surrogate markers with little functional significance, and thus partial overlap with the stem cell populations and sometimes other closely related cells that are not stem cells. Also, some of these markers (e.g., Thy1) are not conserved across mouse species, and use of markers like CD34$^-$ for HSC purification requires mice to be at least 8 weeks old.

SLAM Code

Alternative methods that could give rise to a similar or better harvest of stem cells is an active area of research, and are presently emerging. One such method uses a signature of *SLAM* family of cell surface molecules. The SLAM (Signaling lymphocyte activation molecule) family is a group of more than 10 molecules whose genes are located mostly tandemly in a single locus on chromosome 1 (mouse), all belonging to a subset of the immunoglobulin gene superfamily, and originally thought to be involved in T-cell stimulation. This family includes CD48, CD150, CD244, etc., CD150 being the founding member, and, thus, also known as slamF1, i.e., SLAM family member 1.

The signature SLAM codes for the hemopoietic hierarchy are:

- Hematopoietic stem cells (HSC): CD150+CD48−CD244−

- Multipotent progenitor cells (MPPs): CD150−CD48−CD244+

- Lineage-restricted progenitor cells (LRPs): CD150−CD48+CD244+

- Common myeloid progenitor (CMP): lin−SCA-1−c-kit+CD34+CD16/32mid

- Granulocyte-macrophage progenitor (GMP): lin−SCA-1−c-kit+CD34+C-D16/32hi

- Megakaryocyte-erythroid progenitor (MEP): lin−SCA-1−c-kit+CD34−CD16/32low

For HSCs, CD150^{+CD48-} was sufficient instead of CD150$^{+CD48-CD244-}$ because CD48 is a ligand for CD244, and both would be positive only in the activated lineage-restricted progenitors. It seems that this code was more efficient than the more tedious earlier set of the large number of markers, and are also conserved across the mouse strains; however, recent work has shown that this method excludes a large number of HSCs and includes an equally large number of non-stem cells. $CD150^{+CD48-}$ gave stem cell purity comparable to $Thy1^{loSCA-1+}lin^{-c-kit+}$ in mice.

LT-HSC/ST-HSC/Early MPP/Late MPP

Irving Weissman's group at Stanford University was the first to isolate mouse hematopoietic stem cells in 1988 and was also the first to work out the markers to distinguish the mouse long-term (LT-HSC) and short-term (ST-HSC) hematopoietic stem cells (self-renew-capable), and the Multipotent progenitors (MPP, low or no self-renew capability — the later the developmental stage of MPP, the lesser the self-renewal ability and the more of some of the markers like CD4 and CD135):

- LT-HSC: $CD34^-$, $CD38^-$, $SCA-1^+$, $Thy1.1^{+/lo}$, $C-kit^+$, lin^-, $CD135^-$, $Slamf1/CD150^+$

- ST-HSC: $CD34^+$, $CD38^+$, $SCA-1^+$, $Thy1.1^{+/lo}$, $C-kit^+$, lin^-, $CD135^-$, $Slamf1/CD150^+$, Mac-1 $(CD11b)^{lo}$

- Early MPP: $CD34^+$, $SCA-1^+$, $Thy1.1^-$, $C-kit^+$, lin^-, $CD135^+$, $Slamf1/CD150^-$, Mac-1 $(CD11b)^{lo}$, $CD4^{lo}$

- Late MPP: $CD34^+$, $SCA-1^+$, $Thy1.1^-$, $C-kit^+$, lin^-, $CD135^{high}$, $Slamf1/CD150^-$, Mac-1 $(CD11b)^{lo}$, $CD4^{lo}$

Nomenclature of Hematopoietic Colonies and Lineages

Between 1948 and 1950, the Committee for Clarification of the Nomenclature of Cells and Diseases of the Blood and Blood-forming Organs issued reports on the nomenclature of blood cells. An overview of the terminology is shown below, from earliest to final stage of development:

- [root]blast

- pro[root]cyte

- [root]cyte

- meta[root]cyte

- mature cell name

The root for erythrocyte colony-forming units (CFU-E) is "rubri", for granulocyte-monocyte colony-forming units (CFU-GM) is "granulo" or "myelo" and "mono", for lympocyte colony-forming units (CFU-L) is "lympho" and for megakaryocyte colony-forming units (CFU-Meg) is "megakaryo". According to this terminology, the stages of red blood cell formation would be: rubriblast, prorubricyte, rubricyte, metarubricyte, and erythrocyte. However, the following nomenclature seems to be, at present, the most prevalent:

Com-mittee	"lympho"	"rubri"	"granulo" or "myelo"	"mono"	"mega-karyo"
Lineage	Lymphoid	Myeloid	Myeloid	Myeloid	Myeloid
CFU	CFU-L	CFU-GEM-M→CFU-E	CFU-GEMtM→CFU-GM→C-FU-G	CFU-GEM-M→C-FU-GM→C-FU-M	CFU-GEM-M→C-FU-Meg
Process	lymphocyto-poiesis	erythropoi-esis	granulocytopoiesis	monocyto-poiesis	thrombocy-topoiesis
[root] blast	Lymphoblast	Proerythro-blast	Myeloblast	Monoblast	Megakaryo-blast
pro[-root] cyte	Prolympho-cyte	Polychro-matophilic erythrocyte	Promyelocyte	Promonocyte	Promega-karyocyte
[root] cyte	-	Normoblast	Eosino/neutro/basophilic myelocyte		Megakaryo-cyte
meta[-root] cyte	Large lym-phocyte	Reticulocyte	Eosinophilic/neutrophilic/ basophilic metamyelocyte, Eosinophilic/neutrophilic/ basophilic band cell	Early mono-cyte	-
mature cell name	Small lym-phocyte	Erythrocyte	granulocytes (Eosino/neutro/ basophil)	Monocyte	thrombo-cytes (Plate-lets)

Osteoclasts also arise from hemopoietic cells of the monocyte/neutrophil lineage, specifically CFU-GM.

Colony-forming Units

In the context of hematopoietic stem cells, a colony-forming unit is a subtype of HSC. (This sense of the term is different from colony-forming units of microbes, which is a cell counting unit.) There are various kinds of HSC colony-forming units:

- Colony-forming unit–granulocyte-erythrocyte-monocyte-megakaryocyte (CFU-GEMM)
- Colony-forming unit–lymphocyte (CFU-L)
- Colony-forming unit–erythrocyte (CFU-E)
- Colony-forming unit–granulocyte-macrophage (CFU-GM)
- Colony-forming unit–megakaryocyte (CFU-Meg)
- Colony-forming unit–basophil (CFU-B)
- Colony-forming unit–eosinophil (CFU-Eos)

The above CFUs are based on the lineage. Another CFU, the colony-forming unit–spleen (CFU-S), was the basis of an *in vivo* clonal colony formation, which depends on the ability of infused bone marrow cells to give rise to clones of maturing hematopoietic cells in the spleens of irradiated mice after 8 to 12 days. It was used extensively in early studies, but is now considered to measure more mature progenitor or transit-amplifying cells rather than stem cells.

HSC Repopulation Kinetics

Hematopoietic stem cells (HSC) cannot be easily observed directly, and, therefore, their behaviors need to be inferred indirectly. Clonal studies are likely the closest technique for single cell in vivo studies of HSC. Here, sophisticated experimental and statistical methods are used to ascertain that, with a high probability, a single HSC is contained in a transplant administered to a lethally irradiated host. The clonal expansion of this stem cell can then be observed over time by monitoring the percent donor-type cells in blood as the host is reconstituted. The resulting time series is defined as the repopulation kinetic of the HSC.

The reconstitution kinetics are very heterogeneous. However, using symbolic dynamics, one can show that they fall into a limited number of classes. To prove this, several hundred experimental repopulation kinetics from clonal Thy-1^{lo} SCA-1^+ lin$^-$ c-kit$^+$ HSC were translated into symbolic sequences by assigning the symbols "+", "-", "~" whenever two successive measurements of the percent donor-type cells have a positive, negative, or unchanged slope, respectively. By using the Hamming distance, the repopulation patterns were subjected to cluster analysis yielding 16 distinct groups of kinetics. To finish the empirical proof, the Laplace add-one approach was used to determine that the probability of finding kinetics not contained in these 16 groups is very small. By corollary, this result shows that the hematopoietic stem cell compartment is also heterogeneous by dynamical criteria.

DNA Damage and Aging

DNA strand breaks accumulate in long term HSCs during aging. This accumulation is associated with a broad attenuation of DNA repair and response pathways that depends on HSC quiescence. Non-homologous end joining (NHEJ) is a pathway that repairs double-strand breaks in DNA. NHEJ is referred to as "non-homologous" because the break ends are directly ligated without the need for a homologous template. The NHEJ pathway depends on several proteins including ligase 4, DNA polymerase mu and NHEJ factor 1 (NHEJ1, also known as Cernunnos or XLF).

DNA ligase 4 (Lig4) has a highly specific role in the repair of double-strand breaks by NHEJ. Lig4 deficiency in the mouse causes a progressive loss of HSCs during aging. Deficiency of lig4 in pluripotent stem cells results in accumulation of DNA double-strand breaks and enhanced apoptosis.

In polymerase mu mutant mice, hematopoietic cell development is defective in several peripheral and bone marrow cell populations with about a 40% decrease in bone marrow cell number that includes several hematopoietic lineages. Expansion potential of hematopoietic progenitor cells is also reduced. These characteristics correlate with reduced ability to repair double-strand breaks in hematopoietic tissue.

Deficiency of NHEJ factor 1 in mice leads to premature aging of hematopoietic stem cells as indicated by several lines of evidence including evidence that long-term repopulation is defective and worsens over time. Using a human induced pluripotent stem cell model of NHEJ1 deficiency, it was shown that NHEJ1 has an important role in promoting survival of the primitive hematopoietic progenitors. These NHEJ1 deficient cells possess a weak NHEJ1-mediated repair capacity that is apparently incapable of coping with DNA damages induced by physiological stress, normal metabolism, and ionizing radiation.

The sensitivity of haematopoietic stem cells to Lig4, DNA polymerase mu and NHEJ1 deficiency suggests that NHEJ is a key determinant of the ability of stem cells to maintain themselves against physiological stress over time. Rossi et al. found that endogenous DNA damage accumulates with age even in wild type HSCs, and suggested that DNA damage accrual may be an important physiological mechanism of stem cell aging.

Mesenchymal Stem Cell

Mesenchymal stem cells, or MSCs, are multipotent stromal cells that can differentiate into a variety of cell types, including: osteoblasts (bone cells), chondrocytes (cartilage cells), myocytes (muscle cells) and adipocytes (fat cells). This phenomenon has been documented in specific cells and tissues in living animals and their counterparts growing in tissue culture.

Definition

While the terms *mesenchymal stem cell* and *marrow stromal cell* have been used interchangeably, neither term is sufficiently descriptive:

- Mesenchyme is embryonic connective tissue that is derived from the mesoderm and that differentiates into hematopoietic and connective tissue, whereas MSCs do not differentiate into hematopoietic cells.

- Stromal cells are connective tissue cells that form the supportive structure in which the functional cells of the tissue reside. While this is an accurate description for one function of MSCs, the term fails to convey the relatively recently discovered roles of MSCs in the repair of tissue.

- Because the cells, called MSCs by many labs today, can encompass multipotent cells derived from other non-marrow tissues, such as placenta, umbilical cord blood, adipose tissue, adult muscle, corneal stroma or the dental pulp of deciduous baby teeth, yet do not have the capacity to reconstitute an entire organ, the term multipotent stromal cell has been proposed as a better replacement.The youngest, most primitive MSCs can be obtained from the umbilical cord tissue, namely Wharton's jelly and the umbilical cord blood. However the MSCs are found in much higher concentration in the Wharton's jelly compared to the umbilical cord blood, which is a rich source of hematopoietic stem cells. The umbilical cord is easily obtained after the birth of the newborn, is normally thrown away, and poses no risk for collection. The umbilical cord MSCs have more primitive properties than other adult MSCs obtained later in life, which might make them a useful source of MSCs for clinical applications.

An extremely rich source for mesenchymal stem cells is the developing tooth bud of the mandibular third molar. While considered multipotent, they may prove to be pluripotent. The stem cells eventually form enamel, dentin, blood vessels, dental pulp, and nervous tissues, including a minimum of 29 different unique end organs. Because of extreme ease in collection at 8–10 years of age before calcification, and minimal to no morbidity, they will probably constitute a major source for personal banking, research, and multiple therapies. These stem cells have been shown capable of producing hepatocytes.

Additionally, amniotic fluid has been shown to be a rich source of stem cells. As many as 1 in 100 cells collected during amniocentesis has been shown to be a pluripotent mesenchymal stem cell.

Adipose tissue is one of the richest sources of MSCs. There are more than 500 times more stem cells in 1 gram of fat than in 1 gram of aspirated bone marrow. Adipose stem cells are actively being researched in clinical trials for treatment of a variety of diseases.

The presence of MSCs in peripheral blood has been controversial. However, a few groups have successfully isolated MSCs from human peripheral blood and been able to expand them in culture. Australian company Cynata also claims the ability to mass-produce MSCs from induced pluripotent stem cells obtained from blood cells using the method of K. Hu et al.

Characteristics

Morphology

Mesenchymal stem cells are characterized morphologically by a small cell body with a few cell processes that are long and thin. The cell body contains a large, round nucleus with a prominent nucleolus, which is surrounded by finely dispersed chromatin particles, giving the nucleus a clear appearance. The remainder of the cell body contains a small amount of Golgi apparatus, rough endoplasmic reticulum, mitochondria, and polyribosomes. The cells, which are long and thin, are widely dispersed and the adjacent extracellular matrix is populated by a few reticular fibrils but is devoid of the other types of collagen fibrils.

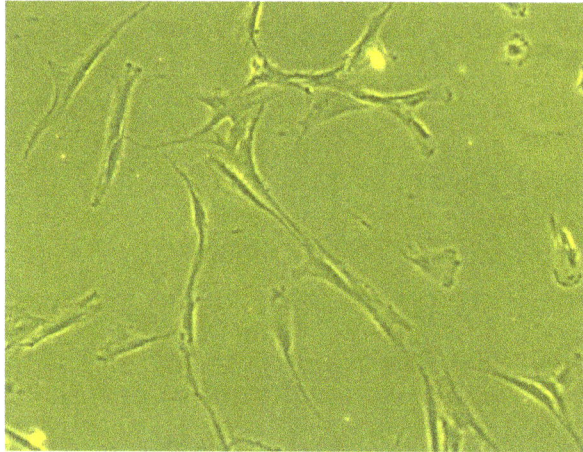

Human bone marrow derived Mesenchymal stem cell showing fibroblast like morphology seen under phase contrast microscope (carl zeiss axiovert 40 CFL) at 63 x magnification

Detection

The International Society for Cellular Therapy (ISCT) has proposed a set of standards to define MSCs. A cell can be classified as an MSC if it shows plastic adherent properties under normal culture conditions and has a fibroblast-like morphology. In fact, some argue that MSCs and fibroblasts are functionally identical. Furthermore, MSCs can undergo osteogenic, adipogenic and chondrogenic differentiation ex-vivo. The cultured MSCs also express on their surface CD73, CD90 and CD105, while lacking the expression of CD11b, CD14, CD19, CD34, CD45, CD79a and HLA-DR surface markers.

Differentiation Capacity

MSCs have a great capacity for self-renewal while maintaining their multipotency. Beyond that, there is little that can be definitively said. The standard test to confirm multipotency is differentiation of the cells into osteoblasts, adipocytes, and chondrocytes as well as myocytes and neurons. MSCs have been seen to even differentiate into neuron-like cells, but there is lingering doubt whether the MSC-derived neurons are functional. The degree to which the culture will differentiate varies among individuals and how differentiation is induced, e.g., chemical vs. mechanical; and it is not clear whether this variation is due to a different amount of "true" progenitor cells in the culture or variable differentiation capacities of individuals' progenitors. The capacity of cells to proliferate and differentiate is known to decrease with the age of the donor, as well as the time in culture. Likewise, whether this is due to a decrease in the number of MSCs or a change to the existing MSCs is not known.Immunomodulatory effects

Numerous studies have demonstrated that human MSCs avoid allorecognition, interfere with dendritic cell and T-cell function, and generate a local immunosuppressive microenvironment by secreting cytokines. It has also been shown that the immunomodulatory function of human MSC is enhanced when the cells are exposed to an inflammatory environment characterised by the presence of elevated local interferon-gamma levels. Other studies contradict some of these findings, reflecting both the highly heterogeneous nature of MSC isolates and the considerable differences between isolates generated by the many different methods under development.

Culturing

The majority of modern culture techniques still take a colony-forming unit-fibroblasts (CFU-F) approach, where raw unpurified bone marrow or ficoll-purified bone marrow Mononuclear cell are plated directly into cell culture plates or flasks. Mesenchymal stem cells, but not red blood cells or haematopoetic progenitors, are adherent to tissue culture plastic within 24 to 48 hours. However, at least one publication has identified a population of non-adherent MSCs that are not obtained by the direct-plating technique.

Other flow cytometry-based methods allow the sorting of bone marrow cells for specific surface markers, such as STRO-1. STRO-1+ cells are generally more homogenous, and have higher rates of adherence and higher rates of proliferation, but the exact differences between STRO-1+ cells and MSCs are not clear.

Methods of immunodepletion using such techniques as MACS have also been used in the negative selection of MSCs.

The supplementation of basal media with fetal bovine serum or human platelet lysate is common in MSC culture. Prior the use of platelet lysates for MSC culture, the pathogen inactivation process is recommended to prevent pathogen transmission.

Cancer

Mesenchymal stem cells have been shown to contribute to cancer progression in a number of different cancers, particularly the hematological malignancies because they contact the transformed blood cells in the bone marrow.

Medical Use

Typical gross appearance of a tubular cartilaginous construct engineered from amniotic mesenchymal stem cells

The mesenchymal stem cells can be activated and mobilized if needed. However, the efficiency is very low. For instance, damage to muscles heals very slowly but further study into mechanisms of MSC action may provide avenues for increasing their capacity for tissue repair.

Many of the early clinical successes using intravenous transplantation have come in systemic diseases like graft versus host disease and sepsis. However, it is becoming more accepted that diseases involving peripheral tissues, such as inflammatory bowel disease, may be better treated with methods that increase the local concentration of cells. Direct injection or placement of cells into a site in need of repair may be the preferred method of treatment, as vascular delivery suffers from a "pulmonary first pass effect" where intravenous injected cells are sequestered in the lungs. Clinical case reports in orthopedic applications have been published, though the number of patients treated is small and these methods still lack rigorous study demonstrating effectiveness. Wakitani has published a small case series of nine defects in five knees involving surgical transplantation of mesenchymal stem cells with coverage of the treated chondral defects.

In Treating Autoimmune Disease

At least 218 clinical trials investigating the efficacy of mesenchymal stem cells in treating diseases have been initiated - many of which being autoimmune diseases. Promising results have been shown in a variety of conditions, such as graft versus host disease, Crohn's disease, multiple sclerosis, systemic lupus erythematosus, and systemic sclerosis. While their anti-inflammatory/immunomodulatory effects appear to greatly ameliorate autoimmune disease severity, the durability of these effects remain to be seen.

Clinical Trials of Cryopreserved MSCs

Scientists have reported that MSCs when transfused immediately within few hours post thawing may show reduced function or show decreased efficacy in treating diseases as compared to those MSCs which are in log phase of cell growth, so cryopreserved MSCs should be brought back into log phase of cell growth in *in vitro* culture before these are administered for clinical trials or experimental therapies, re-culturing of MSCs will help in recovering from the shock the cells get during freezing and thawing. Various clinical trials on MSCs have failed which used cryopreserved product immediately post thaw as compared to those clinical trials which used fresh MSCs.

History

In 1924, Russian-born morphologist Alexander A. Maximow used extensive histological findings to identify a singular type of precursor cell within mesenchyme that develops into different types of blood cells.

Scientists Ernest A. McCulloch and James E. Till first revealed the clonal nature of marrow cells in the 1960s. An *ex vivo* assay for examining the clonogenic potential of multipotent marrow cells was later reported in the 1970s by Friedenstein and colleagues. In this assay system, stromal cells were referred to as colony-forming unit-fibroblasts (CFU-f).

The first clinical trials of MSCs were completed in 1995 when a group of 15 patients were injected with cultured MSCs to test the safety of the treatment. Since then, over 200 clinical trials have been started. However, most are still in the safety stage of testing.

Subsequent experimentation revealed the plasticity of marrow cells and how their fate could be determined by environmental cues. Culturing marrow stromal cells in the presence of osteogenic stimuli such as *ascorbic acid, inorganic phosphate*, and *dexamethasone* could promote their differentiation into osteoblasts. In contrast, the addition of *transforming growth factor-beta* (TGF-b) could induce chondrogenic markers.

Application in Therapy

Statistical-based analysis of MSC therapy for osteo-diseases inferred that most studies are still under investigation. There are different follow-up times that indicate we are still far from reaching the final conclusion.

Amniotic Stem Cells

Amniotic stem cells are the mixture of stem cells that can be obtained from the amniotic fluid as well as the amniotic membrane. They can develop into various tissue types including skin, cartilage, cardiac tissue, nerves, muscle, and bone. The cells also have potential medical applications, especially in organ regeneration.

The stem cells are usually extracted from the amniotic sac by amniocentesis which occurs without harming the embryos. The use of amniotic fluid stem cells is therefore generally considered to lack the ethical problems associated with the use of cells from embryos.

In 2009, the first US amniotic stem cell bank was opened in Medford, MA, by Biocell Center, an international company specializing in the cryopreservation and private banking of amniotic fluid stem cells.

History

The presence of embryonic and foetal cells from all germ layers in the amniotic fluid was gradually determined since the 1980s. Haematopoietic progenitor cells were first reported to be present in the amniotic fluid in 1993, specifically up to the 12th week of pregnancy. It was suggested that these originated from the yolk sac.

In 1996, a study indicated that non-haematopoietic progenitor cells were also present in the amniotic fluid. This was later confirmed as mesenchymal stem cells were obtained. In addition, evidence indicated that embryonic stem cells are part of the fluid, although in very small quantities.

At around the same time, it was determined that stem cells from the amniotic membrane also have multipotent potential. AS their differentiation into neural and glial cells as well as hepatocyte precursors was observed.

Properties

The majority of stem cells present in the amniotic fluid share many characteristics, which suggests they may have a common origin.

In 2007, it was confirmed that the amniotic fluid contains a heterogeneous mixture of multipotent cells after it was demonstrated that they were able to differentiate into cells from all three germ layers but they could not form teratomas following implantation into immunodeficient mice. This characteristic differentiates them from embryonic stem cells but indicates similarities with adult stem cells. However, foetal stem cells attained from the amniotic fluid are more stable and more plastic than their adult counterparts making it easier for them to be reprogrammed to a pluripotent state.

A variety of techniques has been developed for the isolation and culturing of amniotic stem cells. One of the more common isolation methods involves the removal of amniotic fluid by amniocentesis. The cells are then extracted from the fluid based on the presence of c-Kit. Several variations of this method exist. There is some debate whether c-Kit is a suitable marker to distinguish amniotic stem cells from other cell types because cells lacking c-Kit also display differentiation potential. Culture conditions may also be adjusted to promote the growth of a particular cell type.

Mesenchymal Stem Cells

Mesenchymal stem cells (MSCs) are highly abundant in the amniotic fluid and several techniques have been described for their isolation. They usually involve the removal of amniotic fluid by amniocentesis and their distinction from other cells may be based on their morphology or other characteristics.

Human leukocyte antigen testing has been utilised to confirm that the MSCs stem from the fetus and not from the mother. Originally it was proposed that the MSCs were discarded from the embryo at the end of their life cycle but since the cells remained viable in the amniotic fluid and were able to proliferate in culture this hypothesis was overturned. However, it remains unclear whether the cells originate from the fetus itself, the placenta or possibly the inner cell mass of the blastocyst.

Comparison of amniotic fluid-derived MSCs to bone-marrow-derived ones showed that the former has a higher expansion potential in culture. However, the cultured amniotic fluid-derived MSCs have a similar phenotype to both adult bone-marrow-derived MSCs and MSCs originating from second trimester fetal tissue. In animals, the MSCs seem to have a unique immunological profile which was observed after their isolation and *in vitro* culturing.

Embryonic-like Stem Cells

As opposed to mesenchymal stem cells, embryonic-like stem cells are not abundant in the amniotic fluid, making up less than 1% of amniocentesis samples. Embryonic-like stem cells were originally identified using markers common to embryonic stem cells such as nuclear Oct4, CD34, vimentin, alkaline phosphatase, stem cell factor and c-Kit. However, these markers were not necessarily concomitantly expressed. In addition, all of these markers can occur on their own or in some combination in other types of cells.

The pluripotency of these embryonic-like stem cells remains to be fully established. Although those cells which expressed the markers were able to differentiate into muscle, adipogenic, osteogenic, nephrogenic, neural and endothelial cells, this did not necessarily occur from a homogenous population of undifferentiated cells. Evidence in favour of their embryonic stem cell nature is the cells' ability to produce clones.

Clinical Applications

The use of amniotic stem cells instead of embryonic stem cells may be advantageous in some cases for various reasons including that the former do not form teratomas. Their enhanced stability and plasticity compared to adult stem cells may also be an advantage. Stem cells from both the amniotic fluid and membrane are utilised for therapeutic approaches.

Foetal Tissue Engineering

Possible applications include the use of amniotic stem cells for foetal tissue engineering to reconstruct birth defects in infants. This would circumvent the complications that are often associated with harvesting stem cells from foetal tissue. A small amount of amniotic fluid provides a large enough quantity of cells for the tissue engineering process and could help correct a number of defects including diaphragmatic hernia and possibly repair premature membrane rupture during pregnancy. If frozen and banked, the cells may also be used for similar purpose later in life.

Cardiovascular Tissue Engineering

Several studies have been carried out to investigate the potential of amniotic stem cells to differentiate into cardiac cells. Although c-Kit sorted cells express some genes common in cardiac cells, success in this area is still limited. Co-culturing, i.e. mixing cells and plating them together, of human amniotic stem cells with neonatal rat ventricular myocytes (NRVM) caused the cells to form functional gap junctions with each other, an indicator for cardiac-like cells. However, these results may be due to the specific features of the NRVM or fusion of the cells rather than the amniotic stem cell's own potential to differentiate into cardiac cells. In general, these types of techniques are considered to be potentially significant but further investigations are required.

Another area of interest is the use of these cells for improvement of cardiac tissue following a myocardial infarction. Several strategies have been tested in rats including the injection of dissociated amniotic stem cells into the infarct region, which yielded conflicting results from several research groups. In contrast, injection of amniotic stem cell aggregates seems to improve the function of the tissue significantly by reducing the size of the infarct area and improving the function of the left ventricle. In addition, vasculature density has been shown to increase. Injection of cells immediately following the infarct is particularly beneficial as the cells protect the cardiac tissue from further damage.

Kidney Injury Repair

Following the discovery that amniotic stem cells are able to differentiate into renal cells, this was further explored in several studies. These showed that *in vitro* the cells were able to contribute to early kidney structures as well as being able to integrate into early kidney structures *ex vivo* and continue their development into mature nephrons. Results obtained for the use of amniotic stem cells in the postnatal kidney were far less encouraging as the cell's contribution to the tissue was very small. However, the cells were able to exert a protective effect on tubular cells in mice with acute tubular necrosis.

Amniotic stem cells can also be used to treat chronic damage. This was shown in mouse models for Alport syndrome, where the cells prolonged survival of the animals by slowing down the progression of the disease. The same effect was observed in mouse models where human amniotic stem cells were used to treat uretral obstruction.

Ethical Considerations

The use of foetal cells has been highly controversial because the tissue is usually obtained from the foetus following induced abortion. In contrast, foetal stem cells in the amniotic fluid can be obtained through routine prenatal testing without the need for abortion or foetal biopsy.

Dental Pulp Stem Cells

Dental pulp stem cells (DPSCs) are stem cells present in the dental pulp, the soft living tissue within teeth. They are multipotent, so they have the potential to differentiate into a variety of cell types. Other sources of dental stem cells are the dental follicle and the developed periodontal ligament.

A subpopulation of dental pulp stem cells has been described as human Immature Dental Pulp Stem Cells (IDPSC). There are various studies where the importance of these cells and their regenerative capacity has been demonstrated. Through the addition of tissue-specific cytokines, differentiated cells were obtained in vitro from these cells, not only of mesenchymal linage but also of endodermal and ectodermal linage. Among them are the IPS, MAPCs cells.

Several publications have stressed the importance of the expression of pluripotentiality associated markers: the transcription factors Nanog, SOX2, Oct3/4, SSEA4, CD13, are indispensable for the stem cells to divide indefinitely without affecting their differentiation potential, i.e., maintaining their self-renovation capacity. The quantification of protein expression levels in these cells is very important in order to know their pluripotentiality level, as described in some publications.

Atari Metal., established a protocol for isolating and identifying the subpopulations of dental pulp pluripotent-like stem cells (DPPSC). These cells are SSEA4+, OCT3/4+, NANOG+, SOX2+, LIN28+, CD13+, CD105+, CD34-, CD45-, CD90+, CD29+, CD73+, STRO1+ and CD146-, and they show genetic stability in vitro based on genomic analysis with a newly described CGH technique.

DPPSCs were able to form both embryoid body-like structures (EBs) in vitro and teratoma-like structures that contained tissues derived from all three embryonic germ layers when injected in nude mice. DPPSCs can differentiate in vitro into tissues that have similar characteristics to mesoderm, endoderm and ectoderm layers.

Sodium Metaphosphates

Sodium trimetaphosphate and sodium hexametaphosphate have been used to promote the growth, differentiation, and angiogenic potential of HDPCs. Results suggest that these metaphosphates may be candidates for dental pulp tissue engineering and regenerative endodontics.

Definition

Dental pulp is the soft living tissue inside a tooth. Stem cells are found inside the soft living tissue. Scientists have identified the mesenchymal type of stem cell inside dental pulp. This particular type of stem cell has the future potential to differentiate into a variety of other cell types including:

- Myocardiocytes to repair damaged cardiac tissue following a heart attack

- Neuronal to generate nerve and brain tissue

- Myocytes to repair muscle

- Osteocytes to generate bone

- Chondrocytes to generate cartilage

- Adipocytes to generate fat

- Bone and tissue from the oral cavity.

History

- 2005 NIH announces discovery of DPSCs by Dr. Irina Kerkis

- 2006 IDPSC Kerkis reported discovery of Immature Dental Pulp Stem Cells (IDPSC), a pluripotent sub-population of DPSC using dental pulp organ culture.

- 2007 DPSC 1st animal studies begin for bone regeneration.

- 2007 DPSC 1st animal studies begin for dental end uses.

- 2008 DPSC 1st animal studies begin for heart therapies.

- 2008 IDPSC 1st animal study began for muscular dystrophy therapies.

- 2008 DPSC 1st animal studies begin for regenerating brain tissue.

- 2008 DPSC 1st advanced animal study for bone grafting announced. Reconstruction of large size cranial bone defects in rats.

- 2010 IDPSC 1st human trial for cornea replacement

Neural Stem Cell

Neural stem cells (NSCs) are self-renewing, multipotent cells that generate the main phenotype of the nervous system. Stem cells are characterized by their capability to differentiate into multiple cell types via exogenous stimuli from their environment. They undergo asymmetric cell division into two daughter cells, one non-specialized and one specialized. NSCs primarily differentiate into neurons, astrocytes, and oligodendrocytes.

History

In 1989, Sally Temple described multipotent, self-renewing progenitor and stem cells in the subventricular zone (SVZ) of the mouse brain. In 1992, Brent A. Reynolds and Samuel Weiss were the first to isolate neural progenitor and stem cells from the adult striatal tissue, including the SVZ — one of the neurogenic areas — of adult mice brain tissue. In the same year the team of Constance Cepko and Evan Y. Snyder were the first to isolate multipotent cells from the mouse cerebellum and stably transfected them with the oncogene v-myc. Interestingly, this molecule is one of the genes widely used now to reprogram adult non-stem cells into pluripotent stem cells. Since then, neural progenitor and stem cells have been isolated from various areas of the adult brain, including non-neurogenic areas, such as the spinal cord, and from various species including humans.

Aging and Development

In Vivo Origin

There are two basic types of stem cell: adult stem cells, which are limited in their ability to differentiate, and embryonic stem cells (ESCs), which are pluripotent. ESCs are not limited to a particular cell fate; rather they have the capability to differentiate into any cell type. ESCs are derived from the inner cell mass of the blastocyst with the potential to self-replicate.

NSCs are considered adult stem cells because they are limited in their capability to differentiate. NSCs are generated throughout an adult's life via the process of neurogenesis. Since neurons do not divide within the central nervous system (CNS), NSCs can be differentiated to replace lost or injured neurons or in many cases even glial cells. NSCs are differentiated into new neurons within the SVZ of lateral ventricles, a remnant of the embryonic germinal neuroepithelium, as well as the dentate gyrus of the hippocampus.

In Vitro Origin

Adult NSCs were first isolated from mouse striatum in the early 1990s. They are capable of forming multipotent neurospheres when cultured *in vitro*. Neurospheres can produce self-renewing and proliferating specialized cells. These neurospheres can differentiate to form the specified neurons, glial cells, and oligodendrocytes. In previous studies, cultured neurospheres have been transplanted into the brains of immunodeficient neonatal mice and have shown engraftment, proliferation, and neural differentiation.

Communication and Migration

NSCs are stimulated to begin differentiation via exogenous cues from the microenvironment, or stem cell niche. This capability of the NSCs to replace lost or damaged neural cells is called neurogenesis. Some neural cells are migrated from the SVZ along the rostral migratory stream which contains a marrow-like structure with ependymal cells and astrocytes when stimulated. The ependymal cells and astrocytes form glial tubes used by migrating neuroblasts. The astrocytes in the tubes provide support for the migrating cells as well as insulation from electrical and chemical signals released from surrounding cells. The astrocytes are the primary precursors for rapid cell amplification. The neuroblasts form tight chains and migrate towards the specified site of cell damage to repair or replace neural cells. One example is a neuroblast migrating towards the olfactory bulb to differentiate into periglomercular or granule neurons which have a radial migration pattern rather than a tangential one.

On the other hand, the dentate gyrus neural stem cells produce excitatory granule neurons which are involved in learning and memory. One example of learning and memory is pattern separation, a cognitive process used to distinguish similar inputs.

Aging

Neural stem cell proliferation declines as a consequence of aging. Various approaches have been taken to counteract this age-related decline. Because FOXO proteins regulate neural stem cell homeostasis, FOXO proteins have been used to protect neural stem cells by inhibiting Wnt signaling.

Functions During Differentiation and Disease

Epidermal growth factor (EGF) and fibroblast growth factor (FGF) are mitogens that promote neural progenitor and stem cell growth *in vitro*, though other factors synthesized by the neural progenitor and stem cell populations are also required for optimal growth. It is hypothesized that neurogenesis in the adult brain originates from NSCs. The origin and identity of NSCs in the adult brain remain to be defined.

During Differentiation

The most widely accepted model of an adult NSC is a radial, astrocytes-like, GFAP-positive cell. Quiescent stem cells are Type B that are able to remain in the quiescent state due to the renewable tissue provided by the specific niches composed of blood vessels, astrocytes, microglia, ependymal cells, and extracellular matrix present within the brain. These niches provide nourishment, structural support, and protection for the stem cells until they are activated by external stimuli. Once activated, the Type B cells develop into Type C cells, active proliferating intermediate cells, which then divide into neuroblasts consisting of Type A cells. The undifferentiated neuroblasts form chains that migrate and develop into mature neurons. In the olfactory bulb, they mature into GABAergic granule neurons, while in the hippocampus they mature into dentate granule cells.

During Disease

NSCs have an important role during development producing the enormous diversity of neurons, astrocytes and oligodendrocytes in the developing CNS. They also have important role in adult animals, for instance in learning and hippocampal plasticity in the adult mice in addition to supplying neurons to the olfactory bulb in mice.

Notably the role of NSCs during diseases is now being elucidated by several research groups around the world. The responses during stroke, multiple sclerosis, and Parkinson's disease in animal models and humans is part of the current investigation. The results of this ongoing investigation may have future applications to treat human neurological diseases.

Neural stem cells have been shown to engage in migration and replacement of dying neurons in classical experiments performed by Sanjay Magavi and Jeffrey Macklis. Using a laser-induced damage of cortical layers, Magavi showed that SVZ neural progenitors expressing Doublecortin, a critical molecule for migration of neuroblasts, migrated long distances to the area of damage and differentiated into mature neurons expressing NeuN marker. In addition Masato Nakafuku's group from Japan showed for the first time the role of hippocampal stem cells during stroke in mice. These results demonstrated that NSCs can engage in the adult brain as a result of injury. Furthermore, in 2004 Evan Y. Snyder's group showed that NSCs migrate to brain tumors in a directed fashion. Jaime Imitola, M.D and colleagues from Harvard demonstrated for the first time, a molecular mechanism for the responses of NSCs to injury. They showed that chemokines released during injury such as SDF-1a were responsible for the directed migration of human and mouse NSCs to areas of injury in mice. Since then other molecules have been found to participate in the responses of NSCs to injury. All these results have been widely reproduced and expanded by other investigators joining the classical work of Richard L. Sidman in autoradiography to visualize neurogenesis during development, and neurogenesis in the adult by Joseph Altman in the 1960s, as evidence of the responses of adult NSCs activities and neurogenesis during homeostasis and injury.

The search for additional mechanisms that operate in the injury environment and how they influence the responses of NSCs during acute and chronic disease is matter of intense research.

Potential Clinical Applications

Regenerative Therapy of the CNS

Cell death is a characteristic of acute CNS disorders as well as neurodegenerative disease. The loss of cells is amplified by the lack of regenerative abilities for cell replacement and repair in the CNS. One way to circumvent this is to use cell replacement therapy via regenerative NSCs. NSCs can be cultured *in vitro* as neurospheres. These neurospheres are composed of neural stem cells and progenitors (NSPCs) with growth factors such as EGF and FGF. The withdrawal of these growth factors activate differentiation into neurons, astrocytes, or oligodendrocytes which can be transplanted within the brain at the site of injury. The benefits of this therapeutic approach have been examined in Parkinson's disease, Huntington's disease, and multiple sclerosis. NSPCs induce neural repair via intrinsic properties of neuroprotection and immunomodulation. Some possible routes of transplantation include intracerebral transplantation and xenotransplantation.

An alternative therapeutic approach to the transplantation of NSPCs is the pharmacological activation of endogenous NSPCs (eNSPCs). Activated eNSPCs produce neurotrophic factors,several treatments that activate a pathway that involves the phosphorylation of STAT3 on the serine residue and subsequent elevation of Hes3 expression

(STAT3-Ser/Hes3 Signaling Axis) oppose neuronal death and disease progression in models of neurological disorder.

Basic Laboratory Studies

Generation of 3D in Vitro Models of the Human CNS

Human midbrain-derived neural progenitor cells (hmNPCs) have the ability to differentiate down multiple neural cell lineages that lead to neurospheres as well as multiple neural phenotypes. The hmNPC can be used to develop a 3D *in vitro* model of the human CNS. There are two ways to culture the hmNPCs, the adherent monolayer and the neurosphere culture systems. The neurosphere culture system has previously been used to isolate and expand CNS stem cells by its ability to aggregate and proliferate hmNPCs under serum-free media conditions as well as with the presence of epidermal growth factor (EGF) and fibroblast growth factor-2 (FGF2). Initially, the hmNPCs were isolated and expanded before performing a 2D differentiation which was used to produce a single-cell suspension. This single-cell suspension helped achieve a homogenous 3D structure of uniform aggregate size. The 3D aggregation formed neurospheres which was used to form an *in vitro* 3D CNS model.

Bioactive Scaffolds as Traumatic Brain Injury Treatment

Traumatic brain injury (TBI) can deform the brain tissue, leading to necrosis primary damage which can then cascade and activate secondary damage such as excitotoxicity, inflammation, ischemia, and the breakdown of the blood-brain-barrier. Damage can escalate and eventually lead to apoptosis or cell death. Current treatments focus on preventing further damage by stabilizing bleeding, decreasing intracranial pressure and inflammation, and inhibiting pro-apoptoic cascades. In order to repair TBI damage, an upcoming therapeutic option involves the use of NSCs derived from the embryonic peri-ventricular region. Stem cells can be cultured in a favorable 3-dimensional, low cytotoxic environment, a hydrogel, that will increase NSC survival when injected into TBI patients. The intracerebrally injected, primed NSCs were seen to migrate to damaged tissue and differentiate into oligodendrocytes or neuronal cells that secreted neuroprotective factors.

Galectin-1 in Neural Stem Cells

Galectin-1 is expressed in adult NSCs and has been shown to have a physiological role in the treatment of neurological disorders in animal models. There are two approaches to using NSCs as a therapeutic treatment: (1) stimulate intrinsic NSCs to promote proliferation in order to replace injured tissue, and (2) transplant NSCs into the damaged brain area in order to allow the NSCs to restore the tissue. Lentivirus vectors were used to infect human NSCs (hNSCs) with Galectin-1 which were later transplanted into the damaged tissue. The hGal-1-hNSCs induced better and faster brain recovery of the injured tissue as well as a reduction in motor and sensory deficits as compared to only hNSC transplantation.

Assays

Neural stem cells are routinely studied *in vitro* using a method referred to as the Neurosphere Assay (or Neurosphere culture system), first developed by Reynolds and Weiss. Neurospheres are intrinsically heterogeneous cellular entities almost entirely formed by a small fraction (1 to 5%) of slowly dividing neural stem cells and by their progeny, a population of fast-dividing nestin-positive progenitor cells. The total number of these progenitors determines the size of a neurosphere and, as a result, disparities in sphere size within different neurosphere populations may reflect alterations in the proliferation, survival and/or differentiation status of their neural progenitors. Indeed, it has been reported that loss of β1-integrin in a neurosphere culture does not significantly affect the capacity of β1-integrin deficient stem cells to form new neurospheres, but it influences the size of the neurosphere: β1-integrin deficient neurospheres were overall smaller due to increased cell death and reduced proliferation.

While the Neurosphere Assay has been the method of choice for isolation, expansion and even the enumeration of neural stem and progenitor cells, several recent publications have highlighted some of the limitations of the neurosphere culture system as a method for determining neural stem cell frequencies. In collaboration with Reynolds, STEMCELL Technologies has developed a collagen-based assay, called the Neural Colony-Forming Cell (NCFC) Assay, for the quantification of neural stem cells. Importantly, this assay allows discrimination between neural stem and progenitor cells.

Neural Stem Cell Institute

The damaged CNS tissue has very limited regenerative and repair capacity so that loss of neurological function is often chronic and progressive. Cell replacement from stem cells is being actively pursued as a therapeutic option. In 2009, a research institute dedicated solely to translating neural stem research into therapies for patients was created outside of Albany, New York, The Neural Stem Cell Institute.

Neuroepithelial Cell

Neuroepithelial cells are the "stem cells" of the nervous system, deriving from actual stem cells in several different stages of neural development. These neural stem cells then differentiate further into multiple types of cells, like neurons, astrocytes and other glial cells. They appear during embryonic development of the neural tube as well as in adult neurogenesis in specific areas of the central nervous system. They are also associated with several neurodegenerative diseases. These cells have often been called neuroblasts in an effort to delineate them as precursors to neurons and glial cells.

Embryonic Neural Development

Brain Development

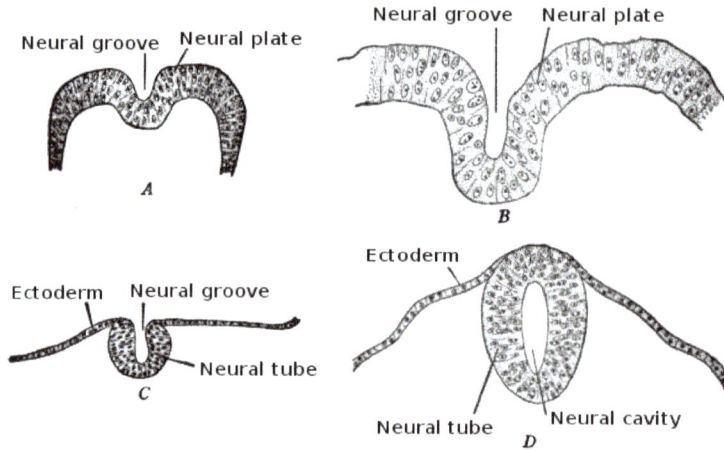

Development of the neural tube

During the third week of embryonic growth the brain begins to develop in the early fetus in a process called induction. Neuroepithelial cells of the ectoderm begin multiplying rapidly and fold in forming the neural plate, which invaginates during the fourth week of embryonic growth and forms the neural tube. The formation of the neural tube polarizes the neuroepithelial cells by orienting the apical side of the cell to face inward, which later becomes the ventricular zone, and the basal side is oriented outward, which contacts the pial, or outer surface of the developing brain. As part of this polarity, neuroepithelial cells express prominin-1 in the apical plasma membrane as well as tight junctions to maintain the cell polarity. Integrin $\alpha6$ anchors the neuroepithelial cells to the basal lamina. The neural tube begins as a single layer of pseudostratified epithelial cells, but rapid proliferation of neuroepithelial cells creates additional layers and eventually three distinct regions of growth. As these additional layers form the apical-basal polarity must be downregulated. Further proliferation of the cells in these regions gives rise to three distinct areas of the brain: the forebrain, midbrain, and hindbrain. The neural plate itself eventually gives rise to the spinal cord.

Neuroepithelial Cell Proliferation

Neuroepithelial cells are a class of stem cell and have similar characteristics, most notably the ability to self renew. During the formation of the neural tube, neuroepithelial cells undergo symmetric proliferative divisions that give rise to two new neuroepithelial cells. At a later stage of brain development, neuroepithelial cells begin to self renew and give rise to non-stem cell progenitors, such as radial glial cells simultaneously by undergoing asymmetric division. Expression of Tis21, an antiproliferative gene, causes the neuroepithelial cell to make the switch from proliferative division to neuronic division. Many of the neuroepithelial cells also divide into radial glial cells, a similar, but more fate restricted cell. Being a more fate restricted cell the radial glial cell will either generate postmitotic neurons, intermediate progenitor cells, or astrocytes in gliogenesis. During neuroepithelial cell division, interkinetic nuclear migration allows the cells to divide unrestricted while maintaining a dense packing. During G1 the cell nucleus migrates to the basal side of the cell and remains there for S phase and migrates to the apical side for G2 phase. This migration requires the help of microtubules and actin filaments.

Radial Glial Cell Transition

Neuroepithelial cells give rise to radial glial cells early on during embryonic development. To make the switch, neuroepithelial cells begin downregulating their epithelial features. During neurulation neuroepithelial cells stop expressing occludin, a tight junction protein. Loss of occludin causes a loss of the previous tight junction seal which is required for the generation of nonepithelial cells such as neurons. Another tight junction protein, PAR3, remains at the apical side of the cell co-localizing with N-cadherin and keeps the apical face of the neuroepithelial cell intact. In the absence of occludin some polarity is still lost and the neuroepithelial cell gives rise to the radial glial cell.

Adult Neurogenesis

Genesis of Neuroepithelial Cells in the Adult CNS

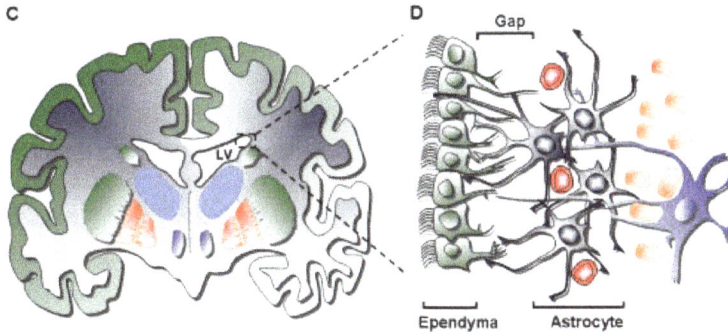

Moving away from the ependymal layer of the SVZ the neural cells become more and more differentiated

In the adult CNS, neuroepithelial cells arise in several different areas of the brain: the subventricular zone (SVZ), the olfactory bulb and the dentate gyrus of the hippocampus. These cells do not appear in any of the peripheral nervous system. Often categorized as neural stem cells, neuroepithelial cells give rise to only a few varieties of neural cells, making them multipotent - a definite distinction from the pluripotent stem cells found in embryonic development. Neuroepithelial cells undergo mitosis generating more neuroepithelial cells, radial glial cells or progenitor cells, the latter two differentiating into either neurons or glial cells. The neuroepithelial cells undergo two different forms of mitosis: asymmetric differentiating division and symmetric prolific division. The asymmetric cell division results in two different varieties of daughter cells (i.e. a neuroepithelial cell divides into a radial glial cell and another neuroepithelial cell), while the symmetric version yields identical daughter cells. This effect is caused by the orientation of the mitotic spindle, which is located in either the posterior or anterior area of the mitotic cell, rather than the center where it is found during symmetric division. The progenitor cells and radial glial cells respond to extracellular trophic factors - like ciliary neurotrophic factor (CNTF), cytokines or neuregulin 1 (NRG1) - that can determine whether the cells will differentiate into either neurons or glia. On a whole, neurogenesis is regulated both by many varied regulatory pathways in the CNS as well as several other factors, from genes to external stimuli such as the individual behavior of a person. The large interconnected web of regulatory responses acts to fine-tune the responses provided by newly formed neurons.

Neurogenesis in Neural Repair

Neurogenesis in the adult brain is often associated with diseases that deteriorate the CNS, like Huntington's disease, Alzheimer's disease, and Parkinson's disease. While adult neurogenesis is up-regulated in the hippocampus in patients with these diseases, whether its effects are regenerative or inconclusive remains to be seen. Individuals with these diseases also often express diminished olfactory abilities as well as decreased cognitive activity in the hippocampus, areas specific to neurogenesis. The genes associated with these diseases like α-synuclein, presenilin 1, MAPT (microtubule associated protein tau) and huntingtin are also often associated with plasticity in the brain and its modification. Neuroplasticity is associated with neurogenesis in a complementary fashion. The new neurons generated by the neuroepithelial cells, progenitors and radial glial cells will not survive unless they are able to integrate into the system by making connections with new neighbors. This also leads to many controversial concepts, like neurogenic therapy involving the transplant of local progenitor cells to a damaged area.

Associated Diseases

Dysembryoplastic Neuroepithelial Tumor (DNT)

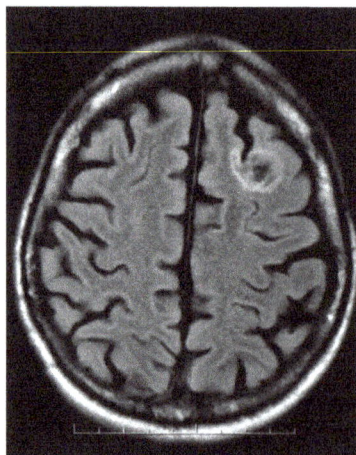

Dysembryoplastic Neuroepithelial Tumor

Dysembryoplastic Neuroepithelial Tumors are a rare, benign tumor that affects children and teenagers under the age of twenty. The tumor occurs in the tissue covering the brain and spinal cord. The symptoms of the tumor are dependent on its location, but most children experience seizures that cannot be controlled by medication. DNT is usually treated through invasive surgery and the patients are usually capable of recovering fully, with little to no long-term effects.

Neuroepithelial Cysts

Neuroepithelial cysts, also known as colloid cysts, develop in individuals between the ages of 20-50 and is relatively rare in individuals under the age of twenty. The cysts are benign tumors that usually appear in the anterior third ventricle. The cysts occur in the epithelium putting their patients at risk for obstructive hydrocephalus, increased intracranial pressure, and rarely intracystic hemorrhage. This results from the cysts enlarging by causing the epithelium to secrete additional mucinous fluid. The cysts are usually found incidentally or if patients become symptomatic presenting with the symptoms of hydrocephalus. The larger cysts are operated on while smaller cysts that are not obstructive can be left alone.

Oligodendroglial Tumors

Oligodendroglial tumors manifest in glial cells, which are responsible for supporting and protecting nerve cells in the brain. The tumor develops over oligodendrocytes and is usually found in the cerebrum around the frontal or temporal lobes. The tumors can either grow slowly in a well-differentiated manner delaying the onset of symptoms, or they can grow rapidly to form an anaplastic oligodendroglioma. The symptoms for this type of tumor include headaches and visual problems. Additionally, blockage of ventricles could cause buildup of cerebral spinal fluid resulting in swelling around the tumor. The location of the tumor may also affect the symptoms since frontal lobe tumors can cause gradual mood or personality changes while temporal lobe tumors result in coordination and speech problems.

Ongoing Research

Neural Chimeras

Researchers have been able to create neural chimeras by combining neurons that developed from embryonic stem cells with glial cells that were also derived from embryonic stem cells. These neural chimeras give researchers a comprehensive way of studying the molecular mechanisms behind cell repair and regeneration via neuroepithelial precursor cells and will hopefully shed light on possible nervous system repair in a clinical setting. In an attempt to identify the key features that differentiate neuroepithelial cells from their progenitor cells, researchers identified an intermediate filament that was expressed by 98% of the neuroepithelial cells of the neural tube, but none of their progenitor cells. After this discovery it became clear that all three cell types in the nervous system resulted from a homogenous population of stem cells. In order make clinical neural repair possible researchers needed to further characterize regional determination of stem cells during brain development by determining what factors commit a precursor to becoming one or the other. While the exact factors that lead to differentiation are unknown, researchers have taken advantage of human-rat neural chimeras to explore the development of human neurons and glial cells in an animal model. These neural chimeras have permitted researchers to look at neurological diseases in an animal model where traumatic and reactive changes can be controlled. Eventually researchers hope to be able to use the information taken from these neural chimera experiments to repair regions of the brain affected by central nervous system disorders. The problem of delivery, however, has still not been resolved as neural chimeras have been shown to circulate throughout the ventricles and incorporate into all parts of the CNS. By finding environmental cues of differentiation, neuroepithelial precursor transplantation could be used in the treatment of many diseases including multiple sclerosis, Huntington's disease, and Parkinson's disease. Further exploration of neural chimera cells and chimeric brains will provide evidence for manipulating the correct genes and increasing the efficacy of neural transplant repair.

Depression

Research on depression indicates that one of the major causal factors of depression, stress, also influences neurogenesis. This connection led researches to postulate that depression could be the result of changes in levels of neurogenesis in the adult brain, specifically in the dentate gyrus. Studies indicate that stress affects neurogenesis by increasing Glucocorticoids and decreasing neurotransmitters such as serotonin. These effects were further verified by inducing stress in lab animals, which resulted in decreased levels of neurogenesis. Additionally, modern therapies that treat depression also promote neurogenesis. Ongoing research is looking to further verify this connection and define the mechanism by which it occurs. This could potentially lead to a better understanding of the development of depression as well as future methods of treatment.

References

- Nakaya, Andrea C. (August 1, 2011). Biomedical ethics. San Diego, CA: ReferencePoint Press. p. 96. ISBN 160152157X.

- Li, Z.; Rana, T. M. (2012). "Using MicroRNAs to Enhance the Generation of Induced Pluripotent Stem Cells". Current Protocols in Stem Cell Biology. doi:10.1002/9780470151808.sc04a04s20. ISBN 0470151803.

- Cibelli, Jose; Lanza, Robert; Campbell, Keith H.S.; West, Michael D. (14 September 2002). Principles of Cloning. Academic Press. ISBN 978-0-08-049215-5.

- GRAHAM, C. F. (January 1977). Teratocarcinoma cells and normal mouse embryogenesiseditor=Michael I. Sherman. MIT Press. ISBN 978-0-262-19158-6.

- ILLMENSEE, K. (14 June 2012). L. B. Russell, ed. Reversion of malignancy and normalized differentiation of teratocarcinoma cells in chimeric mice. Springer London, Limited. pp. 3–24. ISBN 978-1-4684-3392-0.

- Fields,, Mark A.; Hwang,, John; Gong,, Jie; Cai,, Hui; Del Priore, Lucian (9 December 2012). Stephen Tsang, ed. The Eye as a Target Organ for Stem Cell Therapy. Springer. pp. 1–30. ISBN 978-1-4614-5493-9.

- Mikkola, M. (2013) Human pluripotent stem cells: glycomic approaches for culturing and characterization.ISBN 978-952-10-8444-7

- Netter, Frank H. (1987). Musculoskeletal system: anatomy, physiology, and metabolic disorders. Summit, New Jersey: Ciba-Geigy Corporation. p. 134. ISBN 0-914168-88-6.

- Zigova, Tanja; Sanberg, Paul R.; Sanchez-Ramos, Juan Raymond, eds. (2002). Neural stem cells: methods and protocols. Humana Press. ISBN 978-0-89603-964-3. Retrieved 18 April 2010.

- McDonald, A. (2007). "Prenatal Development - The Dana Guide". The Dana Foundation. ISBN 1-932594-10-8. Retrieved 7 December 2011.

- Hodge, Russ (2016-01-25). "Hacking the programs of cancer stem cells". medicalxpress.com. Medical Express. Retrieved 2016-02-12.

- BAERTSCHI, BERNARD, and ALEXANDRE MAURON. "Moral Status Revisited: The Challenge Of Reversed Potency." Bioethics 24.2 (2010): 96-103. Retrieved. 19 Apr. 2015.

- "Stem Cell Basics: Introduction" Bethesda, MD: National Institutes of Health (NIH), U.S. Department of Health and Human Services, 2009. Retrieved. 19 Apr. 2015

- Robertson, John A. "Embryo Stem Cell Research: Ten Years Of Controversy." Journal Of Law, Medicine & Ethics 38.2 (2010): 191-203. Retrieved. 19 Apr. 2015.

3

Stem Cell Therapy

Stem-cell therapies are used to treat diseases. One of the most common forms of stem cell therapy is bone marrow transplant. In recent times, a lot of focus is put on the application of stem cell treatment for diseases such as diabetes and heart diseases. Some of these are human embryonic stem cells clinical trials and hematopoietic stem cell transplantation. This chapter is an overview of the subject matter incorporating all the major aspects of stem cell therapy.

Stem-cell Therapy

Stem-cell therapy is the use of stem cells to treat or prevent a disease or condition.

Bone marrow transplant is the most widely used stem-cell therapy, but some therapies derived from umbilical cord blood are also in use. Research is underway to develop various sources for stem cells, and to apply stem-cell treatments for neurodegenerative diseases and conditions such as diabetes, heart disease, and other conditions.

Stem-cell therapy has become controversial following developments such as the ability of scientists to isolate and culture embryonic stem cells, to create stem cells using somatic cell nuclear transfer and their use of techniques to create induced pluripotent stem cells. This controversy is often related to abortion politics and to human cloning. Additionally, efforts to market treatments based on transplant of stored umbilical cord blood have been controversial.

Medical Uses

For over 30 years, bone marrow has been used to treat cancer patients with conditions such as leukaemia and lymphoma; this is the only form of stem-cell therapy that is widely practiced. During chemotherapy, most growing cells are killed by the cytotoxic agents. These agents, however, cannot discriminate between the leukaemia or neoplastic cells, and the hematopoietic stem cells within the bone marrow. It is this side effect of conventional chemotherapy strategies that the stem-cell transplant attempts to reverse; a donor's healthy bone marrow reintroduces functional stem cells to replace the cells lost in the host's body during treatment. The transplanted cells also generate an immune response that helps to kill off the cancer cells; this process can go too far, however, leading to graft vs host disease, the most serious side effect of this treatment.

Another stem-cell therapy called Prochymal, was conditionally approved in Canada in 2012 for the management of acute graft-vs-host disease in children who are unresponsive to steroids. It is an allogenic stem therapy based on mesenchymal stem cells (MSCs) derived from the bone marrow of adult donors. MSCs are purified from the marrow, cultured and packaged, with up to 10,000 doses derived from a single donor. The doses are stored frozen until needed.

The FDA has approved five hematopoietic stem-cell products derived from umbilical cord blood, for the treatment of blood and immunological diseases.

In 2014, the European Medicines Agency recommended approval of Holoclar, a treatment involving stem cells, for use in the European Union. Holoclar is used for people with severe limbal stem cell deficiency due to burns in the eye.

In March 2016 GlaxoSmithKline's Strimvelis (GSK2696273) therapy for the treatment ADA-SCID was recommended for EU approval.

Research

Stem cells are being studied for a number of reasons. The molecules and exosomes released from stem cells are also being studied in an effort to make medications.

Neurodegeneration

Research has been conducted on the effects of stem cells on animal models of brain degeneration, such as in Parkinson's, Amyotrophic lateral sclerosis, and Alzheimer's disease. There have been preliminary studies related to multiple sclerosis.

Healthy adult brains contain neural stem cells which divide to maintain general stem-cell numbers, or become progenitor cells. In healthy adult laboratory animals, progenitor cells migrate within the brain and function primarily to maintain neuron populations for olfaction (the sense of smell). Pharmacological activation of endogenous neural stem cells has been reported to induce neuroprotection and behavioral recovery in adult rat models of neurological disorder.

Brain and Spinal Cord Injury

Stroke and traumatic brain injury lead to cell death, characterized by a loss of neurons and oligodendrocytes within the brain. A small clinical trial was underway in Scotland in 2013, in which stem cells were injected into the brains of stroke patients.

Clinical and animal studies have been conducted into the use of stem cells in cases of spinal cord injury.

Heart

The pioneering work by Bodo-Eckehard Strauer has now been discredited by the identification of hundreds of factual contradictions. Among several clinical trials that have reported that adult stem-cell therapy is safe and effective, powerful effects have been reported from only a few laboratories, but this has covered old and recent infarcts as well as heart failure not arising from myocardial infarction. While initial animal studies demonstrated remarkable therapeutic effects, later clinical trials achieved only modest, though statistically significant, improvements. Possible reasons for this discrepancy are patient age, timing of treatment and the recent occurrence of a myocardial infarction. It appears that these obstacles may be overcome by additional treatments which increase the effectiveness of the treatment or by optimizing the methodology although these too can be controversial. Current studies vary greatly in cell-procuring techniques, cell types, cell-administration timing and procedures, and studied parameters, making it very difficult to make comparisons. Comparative studies are therefore currently needed.

Stem-cell therapy for treatment of myocardial infarction usually makes use of autologous bone-marrow stem cells (a specific type or all), however other types of adult stem cells may be used, such as adipose-derived stem cells. Adult stem cell therapy for treating heart disease was commercially available in at least five continents as of 2007. Possible mechanisms of recovery include:

- Generation of heart muscle cells

- Stimulation of growth of new blood vessels to repopulate damaged heart tissue

- Secretion of growth factors

- Assistance via some other mechanism

It may be possible to have adult bone-marrow cells differentiate into heart muscle cells.

The first successful integration of human embryonic stem cell derived cardiomyocytes in guinea pigs (mouse hearts beat too fast) was reported in August 2012. The contraction strength was measured four weeks after the guinea pigs underwent simulated heart attacks and cell treatment. The cells contracted synchronously with the existing cells, but it is unknown if the positive results were produced mainly from paracrine as opposed to direct electromechanical effects from the human cells. Future work will focus on how to get the cells to engraft more strongly around the scar tissue. Whether treatments from embryonic or adult bone marrow stem cells will prove more effective remains to be seen.

In 2013 the pioneering reports of powerful beneficial effects of autologous bone marrow stem cells on ventricular function were found to contain "hundreds" of discrepancies. Critics report that of 48 reports there seemed to be just five underlying trials, and that in many cases whether they were randomized or merely observational accepter-versus-rejecter, was contradictory between reports of the same trial. One pair of reports of identical baseline characteristics and final results, was presented in two publications as, respectively, a 578 patient randomized trial and as a 391 patient observational study. Other reports required (impossible) negative standard deviations in subsets of patients, or contained fractional patients, negative NYHA classes. Overall there were many more patients published as having receiving stem cells in trials, than the number of stem cells processed in the hospital's laboratory during that time. A university investigation, closed in 2012 without reporting, was reopened in July 2013.

One of the most promising benefits of stem cell therapy is the potential for cardiac tissue regeneration to reverse the tissue loss underlying the development of heart failure after cardiac injury.

Initially, the observed improvements were attributed to a transdifferentiation of BM-MSCs into cardiomyocyte-like cells. Given the apparent inadequacy of unmodified stem cells for heart tissue regeneration, a more promising modern technique involves treating these cells to create cardiac progenitor cells before implantation to the injured area.

Blood-cell Formation

The specificity of the human immune-cell repertoire is what allows the human body to defend itself from rapidly adapting antigens. However, the immune system is vulnerable to degradation upon the pathogenesis of disease, and because of the critical role that it plays in overall defense, its degradation is often fatal to the organism as a whole. Diseases of hematopoietic cells are diagnosed and classified via a subspecialty of pathology known as hematopathology. The specificity of the immune cells is what allows recognition of foreign antigens, causing further challenges in the treatment of immune disease. Identical matches between donor and recipient must be made for successful transplantation treatments, but matches are uncommon, even between first-degree relatives. Research using both hematopoietic adult stem cells and embryonic stem cells has provided insight into the possible mechanisms and methods of treatment for many of these ailments.Fully mature human red blood cells may be generated *ex vivo* by hematopoietic stem cells (HSCs), which are precursors of red blood cells. In this process, HSCs are grown together with stromal cells, creating an environment that mimics the conditions of bone marrow, the natural site of red-blood-cell growth. Erythropoietin, a growth factor, is added, coaxing the stem cells to complete terminal differentiation into red blood cells. Further research into this technique should have potential benefits to gene therapy, blood transfusion, and topical medicine.

Missing Teeth

In 2004, scientists at King's College London discovered a way to cultivate a complete tooth in mice and were able to grow bioengineered teeth stand-alone in the laboratory. Researchers are confident that the tooth regeneration technology can be used to grow live teeth in human patients.

In theory, stem cells taken from the patient could be coaxed in the lab turning into a tooth bud which, when implanted in the gums, will give rise to a new tooth, and would be expected to be grown in a time over three weeks. It will fuse with the jawbone and release chemicals that encourage nerves and blood vessels to connect with it. The process is similar to what happens when humans grow their original adult teeth. Many challenges remain, however, before stem cells could be a choice for the replacement of missing teeth in the future.

Research is ongoing in different fields, alligators which are polyphyodonts grow up to 50 times a successional tooth (a small replacement tooth) under each mature functional tooth for replacement once a year.

Cochlear Hair Cell Regrowth

Heller has reported success in re-growing cochlea hair cells with the use of embryonic stem cells.

Blindness and Vision Impairment

Since 2003, researchers have successfully transplanted corneal stem cells into damaged eyes to restore vision. "Sheets of retinal cells used by the team are harvested from aborted fetuses, which some people find objectionable." When these sheets are transplanted over the damaged cornea, the stem cells stimulate renewed repair, eventually restore vision. The latest such development was in June 2005, when researchers at the Queen Victoria Hospital of Sussex, England were able to restore the sight of forty patients using the same technique. The group, led by Sheraz Daya, was able to successfully use adult stem cells obtained from the patient, a relative, or even a cadaver. Further rounds of trials are ongoing.

In April 2005, doctors in the UK transplanted corneal stem cells from an organ donor to the cornea of Deborah Catlyn, a woman who was blinded in one eye when acid was thrown in her eye at a nightclub. The cornea, which is the transparent window of the eye, is a particularly suitable site for transplants. In fact, the first successful human transplant was a cornea transplant. The absence of blood vessels within the cornea makes this area a relatively easy target for transplantation. The majority of corneal transplants carried out today are due to a degenerative disease called keratoconus.

The University Hospital of New Jersey reports that the success rate for growth of new cells from transplanted stem cells varies from 25 percent to 70 percent.

In 2014, researchers demonstrated that stem cells collected as biopsies from donor human corneas can prevent scar formation without provoking a rejection response in mice with corneal damage.

In January 2012, The Lancet published a paper by Steven Schwartz, at UCLA's Jules Stein Eye Institute, reporting two women who had gone legally blind from macular degeneration had dramatic improvements in their vision after retinal injections of human embryonic stem cells.

In June 2015, the Stem Cell Ophthalmology Treatment Study (SCOTS), the largest adult stem cell study in ophthalmology (www.clinicaltrials.gov NCT # 01920867) published initial results on a patient with optic nerve disease who improved from 20/2000 to 20/40 following treatment with bone marrow derived stem cells.

Pancreatic Beta Cells

Diabetes patients lose the function of insulin-producing beta cells within the pancreas. In recent experiments, scientists have been able to coax embryonic stem cell to turn into beta cells in the lab. In theory if the beta cell is transplanted successfully, they will be able to replace malfunctioning ones in a diabetic patient.

Transplantation

Human embryonic stem cells may be grown in cell culture and stimulated to form insulin-producing cells that can be transplanted into the patient.

However, clinical success is highly dependent on the development of the following procedures:

- Transplanted cells should proliferate

- Transplanted cells should differentiate in a site-specific manner

- Transplanted cells should survive in the recipient (prevention of transplant rejection)

- Transplanted cells should integrate within the targeted tissue

- Transplanted cells should integrate into the host circuitry and restore function

Orthopaedics

Clinical case reports in the treatment orthopaedic conditions have been reported. To date, the focus in the literature for musculoskeletal care appears to be on mesenchymal stem cells. Centeno et al. have published MRI evidence of increased cartilage and meniscus volume in individual human subjects. The results of trials that include a large number of subjects, are yet to be published. However, a published safety study conducted in a group of 227 patients over a 3-4-year period shows adequate safety and minimal complications associated with mesenchymal cell transplantation.

Wakitani has also published a small case series of nine defects in five knees involving surgical transplantation of mesenchymal stem cells with coverage of the treated chondral defects.

Wound Healing

Stem cells can also be used to stimulate the growth of human tissues. In an adult, wounded tissue is most often replaced by scar tissue, which is characterized in the skin by disorganized collagen structure, loss of hair follicles and irregular vascular structure. In the case of wounded fetal tissue, however, wounded tissue is replaced with normal tissue through the activity of stem cells. A possible method for tissue regeneration in adults is to place adult stem cell "seeds" inside a tissue bed "soil" in a wound bed and allow the stem cells to stimulate differentiation in the tissue bed cells. This method elicits a regenerative response more similar to fetal wound-healing than adult scar tissue formation. Researchers are still investigating different aspects of the "soil" tissue that are conducive to regeneration.

Infertility

Culture of human embryonic stem cells in mitotically inactivated porcine ovarian fibroblasts (POF) causes differentiation into germ cells (precursor cells of oocytes and spermatozoa), as evidenced by gene expression analysis.

Human embryonic stem cells have been stimulated to form Spermatozoon-like cells, yet still slightly damaged or malformed. It could potentially treat azoospermia.

In 2012, oogonial stem cells were isolated from adult mouse and human ovaries and demonstrated to be capable of forming mature oocytes. These cells have the potential to treat infertility.

HIV/AIDS

Destruction of the immune system by the HIV is driven by the loss of CD4+ T cells in the peripheral blood and lymphoid tissues. Viral entry into CD4+ cells is mediated by the interaction with a cellular chemokine receptor, the most common of which are CCR5 and CXCR4. Because subsequent viral replication requires cellular gene expression processes, activated CD4+ cells are the primary targets of productive HIV infection. Recently scientists have been investigating an alternative approach to treating HIV-1/AIDS, based on the creation of a disease-resistant immune system through transplantation of autologous, gene-modified (HIV-1-resistant) hematopoietic stem and progenitor cells (GM-HSPC).

Clinical Trials

GRNOPC1

On 23 January 2009, the US Food and Drug Administration gave clearance to Geron Corporation for the initiation of the first clinical trial of an embryonic stem-cell-based therapy on humans. The trial aimed evaluate the drug GRNOPC1, embryonic stem cell-derived oligodendrocyte progenitor cells, on patients with acute spinal cord injury. The trial was discontinued in November 2011 so that the company could focus on therapies in the "current environment of capital scarcity and uncertain economic conditions". In 2013 biotechnology and regenerative medicine company BioTime (NYSE MKT: BTX) acquired Geron's stem cell assets in a stock transaction, with the aim of restarting the clinical trial.

Cryopreserved Mesenchymal Stromal Cells (MSCs)

Scientists have reported that MSCs when transfused immediately within few hours post thawing may show reduced function or show decreased efficacy in treating diseases as compared to those MSCs which are in log phase of cell growth(fresh), so cryopreserved MSCs should be brought back into log phase of cell growth in invitro culture before these are administered for clinical trials or experimental therapies, re-culturing of MSCs will help in recovering from the shock the cells get during freezing and thawing. Various clinical trials on MSCs have failed which used cryopreserved product immediately post thaw as compared to those clinical trials which used fresh MSCs.

Veterinary Medicine

Research currently conducted on horses, dogs, and cats can benefit the development of stem cell treatments in veterinary medicine and can target a wide range of injuries and diseases such as myocardial infarction, stroke, tendon and ligament damage, osteoarthritis, osteochondrosis and muscular dystrophy both in large animals, as well as humans. While investigation of cell-based therapeutics generally reflects human medical needs, the high degree of frequency and severity of certain injuries in racehorses has put veterinary medicine at the forefront of this novel regenerative approach. Companion animals can serve as clinically relevant models that closely mimic human disease.

Development of Regenerative Treatment Models

Stem cells are thought to mediate repair via five primary mechanisms: 1) providing an anti-inflammatory effect, 2) homing to damaged tissues and recruiting other cells, such as endothelial progenitor cells, that are necessary for tissue growth, 3) supporting tissue remodeling over scar formation, 4) inhibiting apoptosis, and 5) differentiating into bone, cartilage, tendon, and ligament tissue.

To further enrich blood supply to the damaged areas, and consequently promote tissue regeneration, platelet-rich plasma could be used in conjunction with stem cell transplantation. The efficacy of some stem cell populations may also be affected by the method of delivery; for instance, to regenerate bone, stem cells are often introduced in a scaffold where they produce the minerals necessary for generation of functional bone.

Stem cells have also been shown to have a low immunogenicity due to the relatively low number of MHC molecules found on their surface. In addition, they have been found to secrete chemokines that alter the immune response and promote tolerance of the new tissue. This allows for allogeneic treatments to be performed without a high rejection risk.

Sources of Stem Cells

Veterinary applications of stem cell therapy as a means of tissue regeneration have been largely shaped by research that began with the use of adult-derived mesenchymal stem cells to treat animals with injuries or defects affecting bone, cartilage, ligaments and/or tendons. There are two main categories of stem cells used for treatments: allogeneic stem cells derived from a genetically different donor within the same species and autologous mesenchymal stem cells, derived from the patient prior to use in various treatments. A third category, xenogenic stem cells, or stem cells derived from different species, are used primarily for research purposes, especially for human treatments.

Most stem cells intended for regenerative therapy are generally isolated either from the patient's bone marrow or from adipose tissue. Mesenchymal stem cells can differentiate into the cells that make up bone, cartilage, tendons, and ligaments, as well as muscle, neural and other progenitor tissues, they have been the main type of stem cells studied in the treatment of diseases affecting these tissues. The number of stem cells transplanted into damaged tissue may alter efficacy of treatment. Accordingly, stem cells derived from bone marrow aspirates, for instance, are cultured in specialized laboratories for expansion to millions of cells. Although adipose-derived tissue also requires processing prior to use, the culturing methodology for adipose-derived stem cells is not as extensive as that for bone marrow-derived cells. While it is thought that bone-marrow derived stem cells are preferred for bone, cartilage, ligament, and tendon repair, others believe that the less challenging collection techniques and the multi-cellular microenvironment already present in adipose-derived stem cell fractions make the latter the preferred source for autologous transplantation.

New sources of mesenchymal stem cells are being researched, including stem cells present in the skin and dermis which are of interest because of the ease at which they can be harvested with minimal risk to the animal. Hematopoetic stem cells have also been discovered to be travelling in the blood stream and possess equal differentiating ability as other mesenchymal stem cells, again with a very non-invasive harvesting technique.

There has been more recent interest in the use of extra embryonic mesenchymal stem cells. Research is underway to examine the differentiating capabilities of stem cells found in the umbilical cord, yolk sac and placenta of different animals. These stem cells are thought to have more differentiating ability than their adult counterparts, including the ability to more readily form tissues of endodermal and ectodermal origin.

Stem Cells and Hard-tissue Repair

Because of the general positive healing capabilities of stem cells, they have gained interest for the treatment of cutaneous wounds. This is important interest for those with reduced healing capabilities, like diabetics and those undergoing chemotherapy. In one trial, stem cells were isolated from the Wharton's jelly of the umbilical cord. These cells were injected directly into the wounds. Within a week, full re-epithelialization of the wounds had occurred, compared to minor re-epithelialization in the control wounds. This showed the capabilities of mesenchymal stem cells in the repair of epidermal tissues.

Soft-palate defects in horses are caused by a failure of the embryo to fully close at the midline during embryogenesis. These are often not found until after they have become worse because of the difficulty in visualizing the entire soft palate. This lack of visualization is thought to also contribute to the low success rate in surgical intervention to repair the defect. As a result, the horse often has to be euthanized. Recently, the use of mesenchymal stem cells has been added to the conventional treatments. After the surgeon has sutured the palate closed, autologous mesenchymal cells are injected into the soft palate. The stem cells were found to be integrated into the healing tissue especially along the border with the old tissue. There was also a large reduction in the number of inflammatory cells present, which is thought to aid in the healing process.

Stem Cells and Orthopedic Repairs

Autologous stem cell-based treatments for ligament injury, tendon injury, osteoarthritis, osteochondrosis, and sub-chondral bone cysts have been commercially available to practicing veterinarians to treat horses since 2003 in the United States and since 2006 in the United Kingdom. Autologous stem cell based treatments for tendon injury, ligament injury, and osteoarthritis in dogs have been available to veterinarians in the United States since 2005. Over 3000 privately owned horses and dogs have been treated with autologous adipose-derived stem cells. The efficacy of these treatments has been shown in double-blind clinical trials for dogs with osteoarthritis of the hip and elbow and horses with tendon damage.

Tendon Repair

Race horses are especially prone to injuries of the tendon and ligaments. Conventional therapies are very unsuccessful in returning the horse to full functioning potential. Natural healing, guided by the conventional treatments, leads to the formation of fibrous scar tissue that reduces flexibility and full joint movement. Traditional treatments prevented a large number of horses from returning to full activity and also have a high incidence of re-injury due to the stiff nature of the scarred tendon. Introduction of both bone marrow and adipose derived stem cells, along with natural mechanical stimulus promoted the regeneration of tendon tissue. The natural movement promoted the alignment of the new fibers and tendocytes with the natural alignment found in uninjured tendons. Stem cell treatment not only allowed more horses to return to full duty and also greatly reduced the re-injury rate over a three-year period.

The use of embryonic stem cells has also been applied to tendon repair. The embryonic stem cells were shown to have a better survival rate in the tendon as well as better migrating capabilities to reach all areas of damaged tendon. The overall repair quality was also higher, with better tendon architecture and collagen formed. There was also no tumor formation seen during the three month experimental period. Long-term studies need to be carried out to examine the long-term efficacy and risks associated with the use of embryonic stem cells. Similar results have been found in small animals.

Joint Repair

Osteoarthritis is the main cause of joint pain both in animals and humans. Horses and dogs are most frequently affected arthritis. Natural cartilage regeneration is very limited and no current drug therapies are curative, but rather look to reduce the symptoms associated with the degeneration. Different types of mesenchymal stem cells and other additives are still being researched to find the best type of cell and method for long-term treatment.

Adipose-derived mesenchymal cells are currently the most often used because of the non-invasive harvesting. There has been a lot of success recently injecting mesenchymal stem cells directly into the joint. This is a recently developed, non-invasive technique developed for easier clinical use. Dogs receiving this treatment showed greater flexibility in their joints and less pain.

Bone Defect Repair

Bone has a unique and well documented natural healing process that normally is sufficient to repair fractures and other common injuries. Misaligned breaks due to severe trauma, as well as things like tumor resections of bone cancer, are prone to improper healing if left to the natural process alone. Scaffolds composed of natural and artificial components are seeded with mesenchymal stem cells and placed in the defect. Within four weeks of placing the scaffold, newly formed bone begins to integrate with the old bone and within 32 weeks, full union is achieved. Further studies are necessary to fully characterize the use of cell-based therapeutics for treatment of bone fractures.

Stem cells have been used to treat degenerative bone diseases. The normally recommended treatment for dogs that have Legg–Calve–Perthes disease is to remove the head of the femur after the degeneration has progressed. Recently, mesenchymal stem cells have been injected directly in to the head of the femur, with success not only in bone regeneration, but also in pain reduction.

Stem Cells and Muscle Repairs

Stem cells have successfully been used to ameliorate healing in the heart after myocardial infarction in dogs. Adipose and bone marrow derived stem cells were removed and induced to a cardiac cell fate before being injected into the heart. The heart was found to have improved contractility and a reduction in the damaged area four weeks after the stem cells were applied.

A different trial is underway for a patch made of a porous substance on to which the stem cells are "seeded" in order to induce tissue regeneration in heart defects. Tissue was regenerated and the patch was well incorporated into the heart tissue. This is thought to be due, in part, to improved angiogenesis and reduction of inflammation. Although cardiomyocytes were produced from the mesenchymal stem cells, they did not appear to be contractile. Other treatments that induced a cardiac fate in the cells before transplanting had greater success at creating contractile heart tissue.

Stem Cells and Nervous System Repairs

Spinal cord injuries are one of the most common traumas brought into veterinary hospitals. Spinal injuries occur in two ways after the trauma: the primary mechanical damage, and in secondary processes, like inflammation and scar formation, in the days following the trauma. These cells involved in the secondary damage response secrete factors that promote scar formation and inhibit cellular regeneration. Mesenchymal stem cells that are induced to a neural cell fate are loaded on to a porous scaffold and are then implanted at the site of injury. The cells and scaffold secrete factors that counteract those secreted by scar forming cells and promote neural regeneration. Eight weeks later, dogs treated with stem cells showed immense improvement over those treated with conventional therapies. Dogs treated with stem cells were able to occasionally support their own weight, which has not been seen in dogs undergoing conventional therapies.

Treatments are also in clinical trials to repair and regenerate peripheral nerves. Peripheral nerves are more likely to be damaged, but the effects of the damage are not as widespread as seen in injuries to the spinal cord. Treatments are currently in clinical trials to repair severed nerves, with early success. Stem cells induced to a neural fate injected in to a severed nerve. Within four weeks, regeneration of previously damaged stem cells and completely formed nerve bundles were observed.

Stem cells are also in clinical phases for treatment in ophthalmology. Hematopoietic stem cells have been used to treat corneal ulcers of different origin of several horses. These ulcers were resistant to conventional treatments available, but quickly responded positively to the stem cell treatment. Stem cells were also able to restore sight in one eye of a horse with retinal detachment, allowing the horse to return to daily activities.

Keratoconjunctivitis Sicca (KCS)

Pre-clinical models of Sjögrens syndrome have culminated in allogeneic MSCs implanted around the lacrimal glands in KSC dogs that were refractory to current therapy. Significantly improved scores in ocular discharge, conjunctival hyperaemia, corneal changes and Schirmer tear tests (STT) were seen.

Current Areas of Research

Stems Cells in the Lab

The ability to grow up functional adult tissues indefinitely in culture through Directed differentiation creates new opportunities for drug research. Researchers are able to grow up differentiated cell lines and then test new drugs on each cell type to examine possible interactions *in vitro* before performing *in vivo* studies. This is critical in the development of drugs for use in veterinary research because of the possibilities of species specific interactions. The hope is that having these cell lines available for research use will reduce the need for research animals used because effects on human tissue *in vitro* will provide insight not normally known before the animal testing phase.

With the advent of induced pluripotent stem cells (iPSC), treatments being explored and created for the used in endangered low production animals possible. Rather than needing to harvest embryos or eggs, which are limited, the researchers can remove mesenchymal stem cells with greater ease and greatly reducing the danger to the animal due to noninvasive techniques. This allows the limited eggs to be put to use for reproductive purposes only.

Stem Cells and Conservation

Stem cells are being explored for use in conservation efforts. Spermatogonial stem cells have been harvested from a rat and placed into a mouse host and fully mature sperm were produced with the ability to produce viable offspring. Currently research is underway to find suitable hosts for the introduction of donor spermatogonial stem cells. If this becomes a viable option for conservationists, sperm can be produced from high genetic quality individuals who die before reaching sexual maturity, preserving a line that would otherwise be lost.

Future Clinical Uses

The use of stem cells for the treatment of liver disease in both humans and animals has been the focus of considerable interest. The liver has some natural regenerative properties, but is often insufficient to deal with the extent of some liver diseases. Hepatocytes have been formed from some sources of MSC, but they have not been applied clinically currently. There is a large effort to create stem cells differentiated along the pancreatic line as a possible cure for diabetes, but no line has been well established.

Mesenchymal stem cells are currently under clinical trials as a possible treatment for graft v. host disease and graft rejection after experiments on various animals showing that allogenic stem cell treatments were not rejected and showed no difference in healing capabilities compared with autologous stem cells. This is being further researched for creating off-the-shelf allogenic stem cell treatments for various aspects in regenerative veterinary medicine. Clinical trials are underway to explore the low immunogenic properties of stem cells and their possible use for treatment of problems with an overactive immune system seen with allergies and autoimmune disorders.

In recent years, US-based stem-cell clinics have emerged that treat patients with their own bone marrow or adipose derived adult stem cells as part of clinical trials or FDA authorized same day outpatient IRB programs, most notably for athletes to recover from osteoskeletal (bone, joint and connective tissue) related injuries. This emergence of US based human adult stem cell therapy is discussed by Rudderham in his 2012 article Adult Stem Cell US Therapy.

The long-term impact of these treatments will need to be examined outside of their contribution to medicine. Vast improvements in veterinary medicine has allowed for companion and farm animals to live longer lives. This, however, has contributed to the rise in injury and chronic illness in companion animals. Stem cell treatments, especially for the treatment of orthopedic issues in horses, allows for working animals to return to a normal state of activity at a faster rate with a reduction in the re-injury rate.

Embryonic Stem-cell Controversy

There is widespread controversy over the use of human embryonic stem cells. This controversy primarily targets the techniques used to derive new embryonic stem cell lines, which often requires the destruction of the blastocyst. Opposition to the use of human embryonic stem cells in research is often based on philosophical, moral, or religious objections. There is other stem cell research that does not involve the destruction of a human embryo, and such research involves adult stem cells, amniotic stem cells, and induced pluripotent stem cells.

Around the World

China

Stem-cell research and treatment was practiced in the People's Republic of China. The Ministry of Health of the People's Republic of China has permitted the use of stem-cell therapy for conditions beyond those approved of in Western countries. The Western World has scrutinized China for its failed attempts to meet international documentation standards of these trials and procedures.

State-funded companies based in the Shenzhen Hi-Tech Industrial Zone treat the symptoms of numerous disorders with adult stem-cell therapy. Development companies are currently focused on the treatment of neurodegenerative and cardiovascular disorders. The most radical successes of Chinese adult stem cell therapy have been in treating the brain. These therapies administer stem cells directly to the brain of patients with cerebral palsy, Alzheimer's, and brain injuries.Middle East

Since 2008 many universities, centers and doctors tried a diversity of methods; in Lebanon proliferation for stem cell therapy, in-vivo and in-vitro techniques were used, Thus this country is considered the launching place of the **Regentime** procedure. http://www.researchgate.net/publication/281712114_Treatment_of_Long_Standing_Multiple_Sclerosis_with_Regentime_Stem_Cell_Technique The regenerative medicine also took place in Jordan and Egypt.Mexico

Stem-cell treatment is currently being practiced at a clinical level in Mexico. An International Health Department Permit (COFEPRIS) is required. Authorized centers are found in Tijuana, Guadalajara and Cancun. Currently undergoing the approval process is Los Cabos. This permit allows the use of stem cell.South Korea

In 2005, South Korean scientists claimed to have generated stem cells that were tailored to match the recipient. Each of the 11 new stem cell lines was developed using somatic cell nuclear transfer (SCNT) technology. The resultant cells were thought to match the genetic material of the recipient, thus suggesting minimal to no cell rejection.

Thailand

As of 2013, Thailand still considers Hematopoietic stem cell transplants as experimental. Kampon Sriwatanakul began with a clinical trial in October 2013 with 20 patients. 10 are going to receive stem-cell therapy for Type-2 diabetes and the other 10 will receive stem-cell therapy for emphysema. Chotinantakul's research is on Hematopoietic cells and their role for the hematopoietic system function in homeostasis and immune response.

Ukraine

Today, Ukraine is permitted to perform clinical trials of stem-cell treatments (Order of the MH of Ukraine № 630 "About carrying out clinical trials of stem cells", 2008) for the treatment of these pathologies: pancreatic necrosis, cirrhosis, hepatitis, burn disease, diabetes, multiple sclerosis, critical lower limb ischemia. The first medical institution granted the right to conduct clinical trials became the "Institute of Cell Therapy"(Kiev).

Other Countries

Other countries where doctors did stem cells research, trials, manipulation, storage, therapy: Brazil, Cyprus, Germany, Italy, Israel, Japan, Pakistan, Philippines, Russia, Switzerland, Turkey, United Kingdom, India, and many others.

Hematopoietic Stem Cell Transplantation

Hematopoietic stem cell transplantation (HSCT) is the transplantation of multipotent hematopoietic stem cells, usually derived from bone marrow, peripheral blood, or umbilical cord blood. It may be autologous (the patient's own stem cells are used), allogeneic (the stem cells come from a donor) or syngeneic (from an identical twin). It is a medical procedure in the field of hematology, most often performed for patients with certain cancers of the blood or bone marrow, such as multiple myeloma or leukemia. In these cases, the recipient's immune system is usually destroyed with radiation or chemotherapy before the transplantation. Infection and graft-versus-host disease are major complications of allogeneic HSCT.

Hematopoietic stem cell transplantation remains a dangerous procedure with many possible complications; it is reserved for patients with life-threatening diseases. As survival following the procedure has increased, its use has expanded beyond cancer, such as autoimmune diseases.

Medical Uses

Indications

Indications for stem cell transplantation are as follows:

Malignant

- Acute myeloid leukemia (AML)

- Chronic myeloid leukemia (CML)

- Acute lymphoblastic leukemia (ALL)

- Hodgkin lymphoma (relapsed, refractory)

- Non-Hodgkin (relapsed or refractory) lymphoma

- Neuroblastoma

- Ewing sarcoma

- Multiple Myeloma

- Myelodysplastic syndromes

- Gliomas, other solid tumors

Non-malignant

- Thalassemia

- Sickle Cell Anemia

- Aplastic anemia

- Fanconi anemia

- Immune deficiency syndromes

- Inborn errors of metabolism

Many recipients of HSCTs are multiple myeloma or leukemia patients who would not benefit from prolonged treatment with, or are already resistant to, chemotherapy. Candidates for HSCTs include pediatric cases where the patient has an inborn defect such as severe combined immunodeficiency or congenital neutropenia with defective stem cells, and also children or adults with aplastic anemia who have lost their stem cells after birth. Other conditions treated with stem cell transplants include sickle-cell disease, myelodysplastic syndrome, neuroblastoma, lymphoma, Ewing's sarcoma, desmoplastic small round cell tumor, chronic granulomatous disease and Hodgkin's disease. More recently non-myeloablative, "mini transplant(microtransplantation)," procedures have been developed that require smaller doses of preparative chemo and radiation. This has allowed HSCT to be conducted in the elderly and other patients who would otherwise be considered too weak to withstand a conventional treatment regimen.

Number of Procedures

In 2006 a total of 50,417 first hematopoietic stem cell transplants were reported as taking place worldwide, according to a global survey of 1327 centers in 71 countries conducted by the Worldwide Network for Blood and Marrow Transplantation. Of these, 28,901 (57 percent) were autologous and 21,516 (43 percent) were allogeneic (11,928 from family donors and 9,588 from unrelated donors). The main indications for transplant were lymphoproliferative disorders (54.5 percent) and leukemias (33.8 percent), and the majority took place in either Europe (48 percent) or the Americas (36 percent).

In 2014, according to the World Marrow Donor Association, stem cell products provided for unrelated transplantation worldwide had increased to 20,604 (4,149 bone marrow donations, 12,506 peripheral blood stem cell donations, and 3,949 cord blood units).

Graft Types

Autologous

Autologous HSCT requires the extraction (apheresis) of haematopoietic stem cells (HSC) from the patient and storage of the harvested cells in a freezer. The patient is then treated with high-dose chemotherapy with or without radiotherapy with the intention of eradicating the patient's malignant cell population at the cost of partial or complete bone marrow ablation (destruction of patient's bone marrow's ability to grow new blood cells). The patient's own stored stem cells are then transfused into his/her bloodstream, where they replace destroyed tissue and resume the patient's normal blood cell production. Autologous transplants have the advantage of lower risk of infection during the immune-compromised portion of the treatment since the recovery of immune function is rapid. Also, the incidence of patients experiencing rejection (and graft-versus-host disease is impossible) is very rare due to the donor and recipient being the same individual. These advantages have established autologous HSCT as one of the standard second-line treatments for such diseases as lymphoma.

However, for other cancers such as acute myeloid leukemia, the reduced mortality of the autogenous relative to allogeneic HSCT may be outweighed by an increased likelihood of cancer relapse and related mortality, and therefore the allogeneic treatment may be preferred for those conditions. Researchers have conducted small studies using non-myeloablative hematopoietic stem cell transplantation as a possible treatment for type I (insulin dependent) diabetes in children and adults. Results have been promising; however, as of 2009 it was premature to speculate whether these experiments will lead to effective treatments for diabetes.

Allogeneic

Allogeneics HSCT involves two people: the (healthy) donor and the (patient) recipient. Allogeneic HSC donors must have a tissue (HLA) type that matches the recipient. Matching is performed on the basis of variability at three or more loci of the HLA gene, and a perfect match at these loci is preferred. Even if there is a good match at these critical alleles, the recipient will require immunosuppressive medications to mitigate graft-versus-host disease. Allogeneic transplant donors may be *related* (usually a closely HLA matched sibling), *syngeneic* (a monozygotic or 'identical' twin of the patient - necessarily extremely rare since few patients have an identical twin, but offering a source of perfectly HLA matched stem cells) or *unrelated* (donor who is not related and found to have very close degree of HLA matching). Unrelated donors may be found through a registry of bone marrow donors such as the National Marrow Donor Program. People who would like to be tested for a specific family member or friend without joining any of the bone marrow registry data banks may contact a private HLA testing laboratory and be tested with a mouth swab to see if they are a potential match. A "savior sibling" may be intentionally selected by preimplantation genetic diagnosis in order to match a child both regarding HLA type and being free of any obvious inheritable disorder. Allogeneic transplants are also performed using umbilical cord blood as the source of stem cells. In general, by transfusing healthy stem cells to the recipient's bloodstream to reform a healthy immune system, allogeneic HSCTs appear to improve chances for cure or long-term remission once the immediate transplant-related complications are resolved.

A compatible donor is found by doing additional HLA-testing from the blood of potential donors. The HLA genes fall in two categories (Type I and Type II). In general, mismatches of the Type-I genes (i.e. HLA-A, HLA-B, or HLA-C) increase the risk of graft rejection. A mismatch of an HLA Type II gene (i.e. HLA-DR, or HLA-DQB1) increases the risk of graft-versus-host disease. In addition a genetic mismatch as small as a single DNA base pair is significant so perfect matches require knowledge of the exact DNA sequence of these genes for both donor and recipient. Leading transplant centers currently perform testing for all five of these HLA genes before declaring that a donor and recipient are HLA-identical.

Race and ethnicity are known to play a major role in donor recruitment drives, as members of the same ethnic group are more likely to have matching genes, including the genes for HLA.

As of 2013, there were at least two commercialized allogeneic cell therapies, Prochymal and Cartistem.

Sources and Storage of Cells

To limit the risks of transplanted stem cell rejection or of severe graft-versus-host disease in allogeneic HSCT, the donor should preferably have the same human leukocyte antigens (HLA) as the recipient. About 25 to 30 percent of allogeneic HSCT recipients have an HLA-identical sibling. Even so-called "perfect matches" may have mismatched minor alleles that contribute to graft-versus-host disease.

Bone Marrow

In the ctase of a bone marrow transplant, the HSC are removed from a large bone of the donor, typically the pelvis, through a large needle that reaches the center of the bone. The technique is referred to as a bone marrow harvest and is performed under general anesthesia.

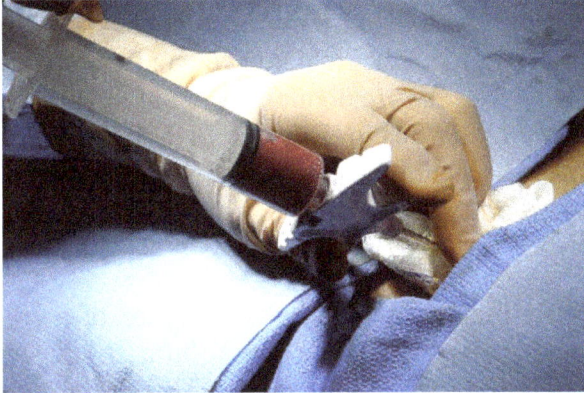

Bone marrow harvest.

Peripheral Blood Stem Cells

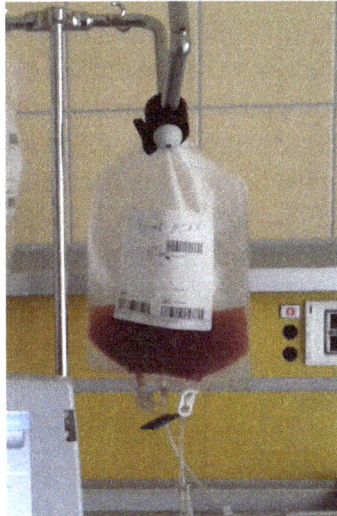

Peripheral blood stem cells

Peripheral blood stem cells are now the most common source of stem cells for HSCT. They are collected from the blood through a process known as apheresis. The donor's blood is withdrawn through a sterile needle in one arm and passed through a machine that removes white blood cells. The red blood cells are returned to the donor. The peripheral stem cell yield is boosted with daily subcutaneous injections of Granulo-cyte-colony stimulating factor, serving to mobilize stem cells from the donor's bone marrow into the peripheral circulation.

Amniotic Fluid

It is also possible to extract stem cells from amniotic fluid for both autologous or heterologous use at the time of childbirth.

Umbilical Cord Blood

Umbilical cord blood is obtained when a mother donates her infant's umbilical cord and placenta after birth. Cord blood has a higher concentration of HSC than is normally found in adult blood. However, the small quantity of blood obtained from an Umbilical Cord (typically about 50 mL) makes it more suitable for transplantation into small children than into adults. Newer techniques using ex-vivo expansion of cord blood units or the use of two cord blood units from different donors allow cord blood transplants to be used in adults.

Cord blood can be harvested from the Umbilical Cord of a child being born after preim-plantation genetic diagnosis (PGD) for human leucocyte antigen (HLA) matching in order to donate to an ill sibling requiring HSCT.

Storage of HSC

Unlike other organs, bone marrow cells can be frozen (cryopreserved) for prolonged periods without damaging too many cells. This is a necessity with autologous HSC because the cells must be harvested from the recipient months in advance of the transplant treatment. In the case of allogeneic transplants, fresh HSC are preferred in order to avoid cell loss that might occur during the freezing and thawing process. Allogeneic cord blood is stored frozen at a cord blood bank because it is only obtainable at the time of childbirth. To cryopreserve HSC, a preservative, DMSO, must be added, and the cells must be cooled very slowly in a controlled-rate freezer to prevent osmotic cellular injury during ice crystal formation. HSC may be stored for years in a *cryofreezer,* which typically uses liquid nitrogen.

Conditioning Regimens

Myeloablative

The chemotherapy or irradiation given immediately prior to a transplant is called the *conditioning regimen*, the purpose of which is to help eradicate the patient's disease prior to the infusion of HSC and to suppress immune reactions. The bone marrow can be *ablated* (destroyed) with dose-levels that cause minimal injury to other tissues. In allogeneic transplants a combination of cyclophosphamide with total body irradiation is conventionally employed. This treatment also has an immunosuppressive effect that prevents rejection of the HSC by the recipient's immune system. The post-transplant prognosis often includes acute and chronic graft-versus-host disease that may be life-threatening. However, in certain leukemias this can coincide with protection against cancer relapse owing to the graft versus tumor effect. *Autologous* transplants may also use similar conditioning regimens, but many other chemotherapy combinations can be used depending on the type of disease.

Non-myeloablative

A newer treatment approach, non-myeloablative allogeneic transplantation, also termed reduced-intensity conditioning (RIC), uses doses of chemotherapy and radiation too low to eradicate all the bone marrow cells of the recipient. Instead, non-myeloablative transplants run lower risks of serious infections and transplant-related mortality while relying upon the *graft versus tumor* effect to resist the inherent increased risk of cancer relapse. Also significantly, while requiring high doses of immunosuppressive agents in the early stages of treatment, these doses are less than for conventional transplants. This leads to a state of mixed chimerism early after transplant where both recipient and donor HSC coexist in the bone marrow space.

Decreasing doses of immunosuppressive therapy then allows donor T-cells to eradicate the remaining recipient HSC and to induce the graft versus tumor effect. This effect is often accompanied by mild graft-versus-host disease, the appearance of which is often a surrogate marker for the emergence of the desirable graft versus tumor effect, and also serves as a signal to establish an appropriate dosage level for sustained treatment with low levels of immunosuppressive agents.

Because of their gentler conditioning regimens, these transplants are associated with a lower risk of transplant-related mortality and therefore allow patients who are considered too high-risk for conventional allogeneic HSCT to undergo potentially curative therapy for their disease. The optimal conditioning strategy for each disease and recipient has not been fully established, but RIC can be used in elderly patients unfit for myeloablative regimens, for whom a higher risk of cancer relapse may be acceptable.

Engraftment

After several weeks of growth in the bone marrow, expansion of HSC and their progeny is sufficient to normalize the blood cell counts and re-initiate the immune system. The offspring of donor-derived hematopoietic stem cells have been documented to populate many different organs of the recipient, including the heart, liver, and muscle, and these cells had been suggested to have the abilities of regenerating injured tissue in these organs. However, recent research has shown that such lineage infidelity does not occur as a normal phenomenon.

Complications

HSCT is associated with a high treatment-related mortality in the recipient (1 percent or higher), which limits its use to conditions that are themselves life-threatening. Major complications are veno-occlusive disease, mucositis, infections (sepsis), graft-versus-host disease and the development of new malignancies.

Infection

Bone marrow transplantation usually requires that the recipient's own bone marrow be destroyed ("myeloablation"). Prior to "engraftment" patients may go for several weeks without appreciable numbers of white blood cells to help fight infection. This puts a patient at high risk of infections, sepsis and septic shock, despite prophylactic antibiotics. However, antiviral medications, such as acyclovir and valacyclovir, are quite effective in prevention of HSCT-related outbreak of herpetic infection in seropositive patients. The immunosuppressive agents employed in allogeneic transplants for the prevention or treatment of graft-versus-host disease further increase the risk of opportunistic infection. Immunosuppressive drugs are given for a minimum of 6-months after a transplantation, or much longer if required for the treatment of graft-versus-host disease. Transplant patients lose their acquired immunity, for example immunity to childhood diseases such as measles or polio. For this reason transplant patients must be re-vaccinated with childhood vaccines once they are off immunosuppressive medications.

Veno-occlusive Disease

Severe liver injury can result from hepatic veno-occlusive disease (VOD). Elevated levels of bilirubin, hepatomegaly and fluid retention are clinical hallmarks of this condition. There is now a greater appreciation of the generalized cellular injury and obstruction in hepatic vein sinuses, and hepatic VOD has lately been referred to as sinusoidal obstruction syndrome (SOS). Severe cases of SOS are associated with a high mortality rate. Anticoagulants or defibrotide may be effective in reducing the severity of VOD but may also increase bleeding complications. Ursodiol has been shown to help prevent VOD, presumably by facilitating the flow of bile.

Mucositis

The injury of the mucosal lining of the mouth and throat is a common regimen-related toxicity following ablative HSCT regimens. It is usually not life-threatening but is very painful, and prevents eating and drinking. Mucositis is treated with pain medications plus intravenous infusions to prevent dehydration and malnutrition.

Graft-versus-host Disease

Graft-versus-host disease (GVHD) is an inflammatory disease that is unique to allogeneic transplantation. It is an attack of the "new" bone marrow's immune cells against the recipient's tissues. This can occur even if the donor and recipient are HLA-identical because the immune system can still recognize other differences between their tissues. It is aptly named graft-versus-host disease because bone marrow transplantation is the only transplant procedure in which the transplanted cells must accept the body rather than the body accepting the new cells.

Acute graft-versus-host disease typically occurs in the first 3 months after transplantation and may involve the skin, intestine, or the liver. High-dose corticosteroids such as prednisone are a standard treatment; however this immuno-suppressive treatment often leads to deadly infections. *Chronic graft-versus-host disease* may also develop after allogeneic transplant. It is the major source of late treatment-related complications, although it less often results in death. In addition to inflammation, chronic graft-versus-host disease may lead to the development of fibrosis, or scar tissue, similar to scleroderma; it may cause functional disability and require prolonged immunosuppressive therapy. Graft-versus-host disease is usually mediated by T cells, which react to foreign peptides presented on the MHC of the host.Graft-versus-tumor effect

Graft versus tumor effect (GVT) or "graft versus leukemia" effect is the beneficial aspect of the Graft-versus-Host phenomenon. For example, HSCT patients with either acute, or in particular chronic, graft-versus-host disease after an allogeneic transplant tend to have a lower risk of cancer relapse. This is due to a therapeutic immune reaction of the grafted donor T lymphocytes against the diseased bone marrow of the recipient. This lower rate of relapse accounts for the increased success rate of allogeneic transplants, compared to transplants from identical twins, and indicates that allogeneic HSCT is a form of immunotherapy. GVT is the major benefit of transplants that do not employ the highest immuno-suppressive regimens.

Graft versus tumor is mainly beneficial in diseases with slow progress, e.g. chronic leukemia, low-grade lymphoma, and some cases multiple myeloma. However, it is less effective in rapidly growing acute leukemias.

If cancer relapses after HSCT, another transplant can be performed, infusing the patient with a greater quantity of donor white blood cells (Donor lymphocyte infusion).

Oral Carcinoma

Patients after HSCT are at a higher risk for oral carcinoma. Post-HSCT oral cancer may have more aggressive behavior with poorer prognosis, when compared to oral cancer in non-HSCT patients.

Prognosis

Prognosis in HSCT varies widely dependent upon disease type, stage, stem cell source, HLA-matched status (for allogeneic HSCT) and conditioning regimen. A transplant offers a chance for cure or long-term remission if the inherent complications of graft versus host disease, immuno-suppressive treatments and the spectrum of opportunistic infections can be survived. In recent years, survival rates have been gradually improving across almost all populations and sub-populations receiving transplants.

Mortality for allogeneic stem cell transplantation can be estimated using the prediction model created by Sorror et al., using the Hematopoietic Cell Transplantation-Specific Comorbidity Index (HCT-CI). The HCT-CI was derived and validated by investigators at the Fred Hutchinson Cancer Research Center (Seattle, WA). The HCT-CI modifies and adds to a well-validated comorbidity index, the Charlson Comorbidity Index (CCI) (Charlson et al.) The CCI was previously applied to patients undergoing allogeneic HCT but appears to provide less survival prediction and discrimination than the HCT-CI scoring system.

Risks to Donor

The risks of a complication depend on patient characteristics, health care providers and the apheresis procedure, and the colony-stimulating factor used (G-CSF). G-CSF drugs include filgrastim (Neupogen, Neulasta), and lenograstim (Graslopin).

Drug Risks

Filgrastim is typically dosed in the 10 microgram/kg level for 4–5 days during the harvesting of stem cells. The documented adverse effects of filgrastim include splenic rupture (indicated by left upper abdominal or shoulder pain, risk 1 in 40000), Adult respiratory distress syndrome (ARDS), alveolar hemorrage, and allergic reactions (usually expressed in first 30 minutes, risk 1 in 300). In addition, platelet and hemoglobin levels dip post-procedure, not returning to normal until one month.

The question of whether geriatrics (patients over 65) react the same as patients under 65 has not been sufficiently examined. Coagulation issues and inflammation of atherosclerotic plaques are known to occur as a result of G-CSF injection. G-CSF has also been described to induce genetic changes in mononuclear cells of normal donors. There is evidence that myelodysplasia (MDS) or acute myeloid leukaemia (AML) can be induced by GCSF in susceptible individuals.

Access Risks

Blood was drawn peripherally in a majority of patients, but a central line to jugular/subclavian/femoral veins may be used in 16 percent of women and 4 percent of men. Adverse reactions during apheresis were experienced in 20 percent of women and 8 percent of men, these adverse events primarily consisted of numbness/tingling, multiple line attempts, and nausea.

Clinical Observations

A study involving 2408 donors (18–60 years) indicated that bone pain (primarily back and hips) as a result of filgrastim treatment is observed in 80 percent of donors by day 4 post-injection. This pain responded to acetaminophen or ibuprofen in 65 percent of donors and was characterized as mild to moderate in 80 percent of donors and severe in 10 percent. Bone pain receded post-donation to 26 percent of patients 2 days post-donation, 6 percent of patients one week post-donation, and <2 percent 1 year post-donation. Donation is not recommended for those with a history of back pain. Other symptoms observed in more than 40 percent of donors include myalgia, headache, fatigue, and insomnia. These symptoms all returned to baseline 1 month post-donation, except for some cases of persistent fatigue in 3 percent of donors.

In one metastudy that incorporated data from 377 donors, 44 percent of patients reported having adverse side effects after peripheral blood HSCT. Side effects included pain prior to the collection procedure as a result of GCSF injections, post-procedural generalized skeletal pain, fatigue and reduced energy.

Severe Reactions

A study that surveyed 2408 donors found that serious adverse events (requiring prolonged hospitalization) occurred in 15 donors (at a rate of 0.6 percent), although none of these events were fatal. Donors were not observed to have higher than normal rates of cancer with up to 4–8 years of follow up. One study based on a survey of medical teams covered approximately 24,000 peripheral blood HSCT cases between 1993 and 2005, and found a serious cardiovascular adverse reaction rate of about 1 in 1500. This study reported a cardiovascular-related fatality risk within the first 30 days HSCT of about 2 in 10000. For this same group, severe cardiovascular events were observed with a rate of about 1 in 1500. The most common severe adverse reactions were pulmonary edema/deep vein thrombosis, splenic rupture, and myocardial infarction. Haematological malignancy induction was comparable to that observed in the general population, with only 15 reported cases within 4 years.

History

Georges Mathé, a French oncologist, performed the first European bone marrow transplant in November 1958 on five Yugoslavian nuclear workers whose own marrow had been damaged by irradiation caused by a criticality accident at the Vinča Nuclear Institute, but all of these transplants were rejected. Mathé later pioneered the use of bone marrow transplants in the treatment of leukemia.

Stem cell transplantation was pioneered using bone-marrow-derived stem cells by a team at the Fred Hutchinson Cancer Research Center from the 1950s through the 1970s led by E. Donnall Thomas, whose work was later recognized with a Nobel Prize in Physiology or Medicine. Thomas' work showed that bone marrow cells infused intravenously could repopulate the bone marrow and produce new blood cells. His work also reduced the likelihood of developing a life-threatening complication called graft-versus-host disease.

The first physician to perform a successful human bone marrow transplant on a disease other than cancer was Robert A. Good at the University of Minnesota in 1968. In 1975, John Kersey, M.D., also of the University of Minnesota, performed the first successful bone marrow transplant to cure lymphoma. His patient, a 16-year-old-boy, is today the longest-living lymphoma transplant survivor.

Donor Registration and Recruitment

At the end of 2012, 20.2 million people had registered their willingness to be a bone marrow donor with one of the 67 registries from 49 countries participating in Bone Marrow Donors Worldwide. 17.9 million of these registered donors had been ABDR typed, allowing easy matching. A further 561,000 cord blood units had been received by one of 46 cord blood banks from 30 countries participating. The highest total number of bone marrow donors registered were those from the USA (8.0 million), and the highest number per capita were those from Cyprus (15.4 percent of the population).

Within the United States, racial minority groups are the least likely to be registered and therefore the least likely to find a potentially life-saving match. In 1990, only six African-Americans were able to find a bone marrow match, and all six had common European genetic signatures.

Africans are more genetically diverse than people of European descent, which means that more registrations are needed to find a match. Bone marrow and cord blood banks exist in South Africa, and a new program is beginning in Nigeria. Many people belonging to different races are requested to donate as there is a shortage of donors in African, Mixed race, Latino, Aboriginal, and many other communities.

Research

HIV

In 2007, a team of doctors in Berlin, Germany, including Gero Hütter, performed a stem cell transplant for leukemia patient Timothy Ray Brown, who was also HIV-positive. From 60 matching donors, they selected a [CCR5]-Δ32 homozygous individual with two genetic copies of a rare variant of a cell surface receptor. This genetic trait confers resistance to HIV infection by blocking attachment of HIV to the cell. Roughly one in 1000 people of European ancestry have this inherited mutation, but it is rarer in other populations. The transplant was repeated a year later after a leukemia relapse. Over three years after the initial transplant, and despite discontinuing antiretroviral therapy, researchers cannot detect HIV in the transplant recipient's blood or in various biopsies of his tissues. Levels of HIV-specific antibodies have also declined, leading to speculation that the patient may have been functionally cured of HIV. However, scientists emphasise that this is an unusual case. Potentially fatal transplant complications (the "Berlin patient" suffered from graft-versus-host disease and leukoencephalopathy) mean that the procedure could not be performed in others with HIV, even if sufficient numbers of suitable donors were found.

In 2012, Daniel Kuritzkes reported results of two stem cell transplants in patients with HIV. They did not, however, use donors with the Δ32 deletion. After their transplant procedures, both were put on antiretroviral therapies, during which neither showed traces of HIV in their blood plasma and purified CD4 T cells using a sensitive culture method (less than 3 copies/mL). However, the virus was once again detected in both patients some time after the discontinuation of therapy.

Multiple Sclerosis

Since McAllister's 1997 report on a patient with multiple sclerosis (MS) who received a bone marrow transplant for CML, over 600 reports have been published describing HSCTs performed primarily for MS. These have been shown to "reduce or eliminate ongoing clinical relapses, halt further progression, and reduce the burden of disability in some patients" that have aggressive highly active MS, "in the absence of chronic treatment with disease-modifying agents".

Clincs performing HSCT includes Northwestern University and Karolinska University Hospital.

Human Embryonic Stem Cells Clinical Trials

The Food and Drug Administration (FDA) approved the first clinical trial in the United States involving human embryonic stem cells on January 23, 2009. Geron Corporation, a biotechnology firm located in Menlo Park, California, originally planned to enroll ten patients suffering from spinal cord injuries to participate in the trial. The company hoped that GRNOPC1, a product derived from human embryonic stem cells, would stimulate nerve growth in patients with debilitating damage to the spinal cord. The trial began in 2010 after being delayed by the FDA because cysts were found on mice injected with these cells, and safety concerns were raised.

FDA Approval Process

In the United States, the FDA must approve all clinical trials involving newly developed pharmaceuticals. Researchers must complete an Investigational New Drug (IND) application in order to earn the FDA's approval. IND applications typically include data from animal and toxicology studies in which the drug's safety is tested, drug manufacturing information explaining how and where the drug will be produced, and a detailed research protocol stating who will be included in the study, how the drug will be administered and how participants will be consented. Testing for new drugs must successfully go through three phases of research before a drug can be marketed to the public. In Phase I trials, the drug's safety is tested on a small group of participants. The drug's effectiveness is tested during Phase II trials with a larger number of participants. Phase III trials, involving 1,000- 3,000 participants, analyze effectiveness, determine side effects and compare the outcomes of the new drug to similar drugs on the market. An additional phase, Phase IV, is included to continually gain information after a drug is on the market. Geron's IND application for the GRNOPC1 clinical trial, nearly 28,000 pages in length, was one of the most extensive applications ever to be submitted to the FDA.

Pre-clinical Data

Before Geron could test GRNOPC1 in humans, tests in animals had to occur. At the University of California at Irvine, Dr. Hans Keirstead and Dr. Gabriel Nistor, credited with the technique used to develop oligodendrocytes from human embryonic stem cells, injected the cells into rats with spinal cord injuries. The condition of the rats improved after treatment.

Geron Spinal Cord Injury Trial

The first patient, identified in an article by the Washington Post as Timothy J. Atchison of Alabama, enrolled in the trial in October 2010. The patient was treated at the Shepherd Center in Atlanta, GA just two weeks after he sustained a spinal cord injury in a car accident. The Shepherd Center and six other spinal centers were recruited by Geron to participate in the clinical trial. The Washington Post reported that Atchison "has begun to get some very slight sensation: He can feel relief when he lifts a bowling ball off his lap and discern discomfort when he pulls on hairs on some parts of his legs. He has also strengthened his abdomen." Atchison underwent therapy at the Shepherd Center for three months before returning home to Alabama.

Although Geron initially aimed to enroll ten patients in the trial, only three additional patients were added after Atchison. As specified by Geron, eligible patients had to experience a neurologically complete spinal cord injury within seven to fourteen days prior to enrollment. In addition, patients had to be between the ages of 18 and 65 and could not have a history of malignancy, significant organ damage, be pregnant or nursing, unable to communicate or participate in any other experimental procedures. Participants received one injection of GRNOPC1 containing approximately 2 million cells. Even though the trial has officially ended, Geron will continue to monitor participants for fifteen years.

Although no official results from the trial have been published, preliminary results from the clinical trial were presented at the American Congress of Rehabilitation Medicine (ACRM) conference in October 2011. None of the participants experienced serious adverse events, although nausea and low magnesium were reported. In addition, no changes to the spinal cord or neurological condition were found.

After investing millions of dollars in the research leading up to this trial, Geron Corporation discontinued the study in November 2011 to focus on cancer research. John Scarlett, Geron's chief executive officer, said "In the current environment of capital scarcity and uncertain economic conditions, we intend to focus our resources on advancing our two novel and promising oncology drug candidates." The company's stocks fell dramatically to $1.50 per share from $2.28 per share when news of the trial's discontinuation became public. A spokesperson for the company said that Geron would save money by ending the trial despite the loss in investors. Because many believed Geron's trial offered hope for advancing knowledge related to stem cells and their potential uses, there was disappointment in the scientific community when the trial was cut short. An article on Bioethics Forum, a publication produced by The Hastings Center, stated, "It is one thing to close a trial to further enrollment for scientific reasons, such as a problem with trial design, or for ethical reasons, such as an unanticipated serious risk of harm to participants. It is quite another matter to close a trial for business reasons, such as to improve profit margins."

In 2013 Geron's stem cell assets were acquired by biotechnology firm BioTime, helmed by CEO Michael D. West, the founder of Geron and former Chief Scientific Officer of Advanced Cell Technology. BioTime indicated that it plans to re-start the embryonic stem cell-based clinical trial for spinal cord injury.

ACT Stargardt's Macular Dystrophy and Dry Age-related Macular Degeneration Clinical Trial

Two clinical trials involving derivatives of human embryonic stem cells were approved in 2010. Advanced Cell Technology (ACT) located in Marlborough, Massachusetts, leads the trials aimed at improving the vision of patients with Stargardt's Macular Dystrophy and Dry Age-Related Macular Degeneration. Originally, twelve patients were estimated to enroll at three hospitals in the U.S.; participating institutions included the Casey Eye Institute in Portland, Oregon, University of Massachusetts Memorial Medical Center in Worchester, Massachusetts, and the New Jersey Medical School in Newark, New Jersey. Patients' eyes were injected with retinal pigmented epithelial cells derived from human embryonic stem cells. While no definitive findings from this study have been produced, an article published in Lancet in January 2012 stated that preliminary findings appear to be promising. In this article, outcomes from two patients treated as part of the trial were discussed. During the trial, neither patients' vision worsened, and no negative side effects were reported.

Phase I/II clinical trials involving retinal pigment epithelial (RPE) cells, derived from human embryonic stem cells, for the treatment of severe myopia were approved in February 2013.

ViaCyte Human Stem Cell-derived Beta Cells for the Treatment of Diabetes Clinical Trial

The FDA approved a phase I clinical trial with ViaCyte beta cells derived from human embryonic stem cell for the treatment of diabetes in August 2014. The cells will be delivered in immunoprotective capsules and pre-clinical results in animal models showed remission of symptoms within a few months. The company reported the successful transplantation of the cells in the first of the 40 patients that will be treated under the trial in late October 2014.

Future Funding

As state funding for human embryonic stem cell research grows, there seems to be more support for state sponsored clinical trials. In 2010, California committed fifty million dollars to early-stage clinical trials. Although approved trials must take place in California, scientists are hopeful that this funding will bolster future research in the field.

References

- Kaushansky, K; Lichtman, M; Beutler, E; Kipps, T; Prchal, J; Seligsohn, U. (2010). Williams Hematology (8th ed.). McGraw-Hill. ISBN 978-0071621519.

- "Human limbal biopsy–derived stromal stem cells prevent corneal scarring". Science Translational Medicine. 12 December 2014. Retrieved 2015-08-02.

- European Medicines Agency. "First stem-cell therapy recommended for approval in EU". Retrieved 12 December 2014.

- Ghosh, Pallab (27 May 2013) Stroke patients see signs of recovery in stem cell trial BBC News health, Retrieved 27 May 2013

- Francis, DP; Mielewczik, M; Zargaran, D; Cole, GD (26 June 2013). "Autologous bone marrow-derived stem cell therapy in heart disease: Discrepancies and contradictions". International Journal of Cardiology. 168 (4): 3381–403. doi:10.1016/j.ijcard.2013.04.152. PMID 23830344. Retrieved 21 July 2013.

- Francis, Darrel P (Oct 2013). "Autologous bone marrow-derived stem cell therapy in heart disease: Discrepancies and contradictions". Int J Cardiol. Elsevier. 168: 3381–403. doi:10.1016/j.ijcard.2013.04.152. PMID 23830344. Retrieved 6 July 2013.

- McNeil, Donald (11 May 2012). "Finding a Match, and a Mission: Helping Blacks Survive Cancer". The New York Times. Retrieved 15 May 2012.

Bone Marrow: An Integrated Study

Bone marrow is the tissue found in the interior of the bones. There are two types of bone marrow, red marrow and yellow marrow. Red blood cells and most of the white blood cells are found in the red marrow. At birth all the bone marrow is red but with age it converts itself into the yellow type. This section will provide an integrated understanding of bone marrow.

Bone Marrow

Bone marrow is the flexible tissue in the interior of bones. In humans, red blood cells are produced by cores of bone marrow in the heads of long bones in a process known as hematopoiesis. On average, bone marrow constitutes 4% of the total body mass of humans; in an adult having 65 kilograms of mass (143 lbs), bone marrow typically accounts for approximately 2.6 kilograms (5.7 lb). The hematopoietic component of bone marrow produces approximately 500 billion blood cells per day, which use the bone marrow vasculature as a conduit to the body's systemic circulation. Bone marrow is also a key component of the lymphatic system, producing the lymphocytes that support the body's immune system.

Bone marrow transplants can be conducted to treat severe diseases of the bone marrow, including certain forms of cancer such as leukemia. Additionally, bone marrow stem cells have been successfully transformed into functional neural cells, and can also potentially be used to treat illnesses such as inflammatory bowel disease.

Structure

Types of Bone Marrow

The two types of bone marrow are "red marrow" (Latin: *medulla ossium rubra*), which consists mainly of hematopoietic tissue, and "yellow marrow" (Latin: *medulla ossium flava*), which is mainly made up of fat cells. Red blood cells, platelets, and most white blood cells arise in red marrow. Both types of bone marrow contain numerous blood vessels and capillaries. At birth, all bone marrow is red. With age, more and more of it is converted to the yellow type; only around half of adult bone marrow is red. Red marrow is found mainly in the flat bones, such as the pelvis, sternum, cranium, ribs, vertebrae and scapulae, and in the cancellous ("spongy") material at the epiphyseal ends of long bones such as the femur and humerus. Yellow marrow is found in the medullary cavity, the hollow interior of the middle portion of short bones. In cases of severe blood loss, the body can convert yellow marrow back to red marrow to increase blood cell production.

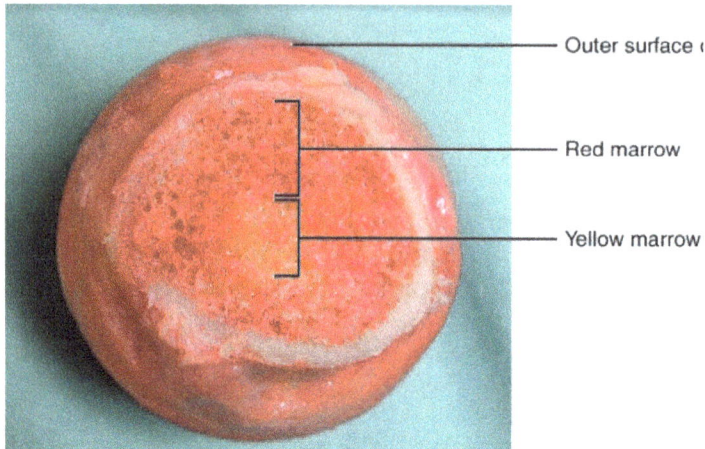

A femoral head with a cortex of bone and medulla of trabecular bone. Both red bone marrow and a central focus of yellow bone marrow are visible.

Stroma

The stroma of the bone marrow is all tissue not directly involved in the marrow's primary function of hematopoiesis. Yellow bone marrow makes up the majority of bone marrow stroma, in addition to smaller concentrations of stromal cells located in the red bone marrow. Though not as active as parenchymal red marrow, stroma is indirectly involved in hematopoiesis, since it provides the hematopoietic microenvironment that facilitates hematopoiesis by the parenchymal cells. For instance, they generate colony stimulating factors, which have a significant effect on hematopoiesis. Cell types that constitute the bone marrow stroma include:

- fibroblasts (reticular connective tissue)

- macrophages, which contribute especially to red blood cell production, as they deliver iron for hemoglobin production.

- adipocytes (fat cells)

- osteoblasts (synthesize bone)

- osteoclasts (resorb bone)

- endothelial cells, which form the sinusoids. These derive from endothelial stem cells, which are also present in the bone marrow.

Cellular Components

In addition, the bone marrow contains hematopoietic stem cells, which give rise to the three classes of blood cells that are found in the circulation: white blood cells (leukocytes), red blood cells (erythrocytes), and platelets (thrombocytes).

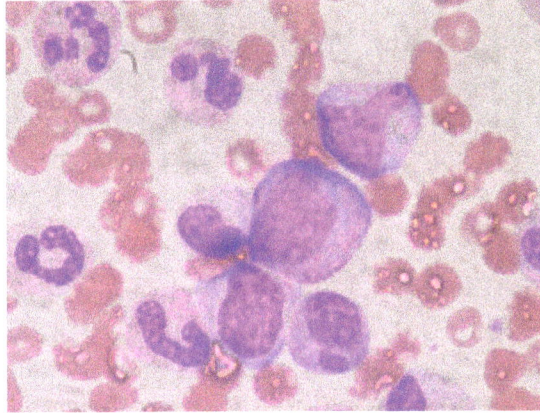

Hematopoietic precursor cells: promyelocyte in the center, two metamyelocytes next to it and band cells from a bone marrow aspirate.

Cellular constitution of the red bone marrow parenchyma			
Group	**Cell type**	**Average fraction**	**Reference range**
Myelopoietic cells	Myeloblasts	0.9%	0.2-1.5
	Promyelocytes	3.3%	2.1-4.1
	Neutrophilic myelocytes	12.7%	8.2-15.7
	Eosinophilic myelocytes	0.8%	0.2-1.3
	Neutrophilic metamyelocytes	15.9%	9.6-24.6
	Eosinophilic metamyelocytes	1.2%	0.4-2.2
	Neutrophilic band cells	12.4%	9.5-15.3
	Eosinophilic band cells	0.9%	0.2-2.4
	Segmented neutrophils	7.4%	6.0-12.0
	Segmented eosinophils	0.5%	0.0-1.3
	Segmented basophils and mast cells	0.1%	0.0-0.2
Erythropoietic cells	Pronormoblasts	0.6%	0.2-1.3
	Basophilic normoblasts	1.4%	0.5-2.4
	Polychromatic normoblasts	21.6%	17.9-29.2
	Orthochromatic normoblast	2.0%	0.4-4.6
Other celltypes	Megakaryocytes	< 0.1%	0.0-0.4
	Plasma cells	1.3%	0.4-3.9
	Reticular cells	0.3%	0.0-0.9
	Lymphocytes	16.2%	11.1-23.2
	Monocytes	0.3%	0.0-0.8

Function

Mesenchymal Stem Cells

The bone marrow stroma contains mesenchymal stem cells (MSCs), also known as marrow stromal cells. These are multipotent stem cells that can differentiate into a variety of cell types. MSCs have been shown to differentiate, in vitro or in vivo, into osteoblasts, chondrocytes, myocytes, adipocytes and beta-pancreatic islets cells.

Bone Marrow Barrier

The blood vessels of the bone marrow constitute a barrier, inhibiting immature blood cells from leaving the marrow. Only mature blood cells contain the membrane proteins, such as aquaporin and glycophorin, that are required to attach to and pass the blood vessel endothelium. Hematopoietic stem cells may also cross the bone marrow barrier, and may thus be harvested from blood.

Lymphatic Role

The red bone marrow is a key element of the lymphatic system, being one of the primary lymphoid organs that generate lymphocytes from immature hematopoietic progenitor cells. The bone marrow and thymus constitute the primary lymphoid tissues involved in the production and early selection of lymphocytes. Furthermore, bone marrow performs a valve-like function to prevent the backflow of lymphatic fluid in the lymphatic system.

Compartmentalization

Biological compartmentalization is evident within the bone marrow, in that certain cell types tend to aggregate in specific areas. For instance, erythrocytes, macrophages, and their precursors tend to gather around blood vessels, while granulocytes gather at the borders of the bone marrow.

Society and Culture

Animal bone marrow has been used in cuisine worldwide for millennia, such as the famed Milanese Ossobuco.Clinical significance

Disease

The normal bone marrow architecture can be damaged or displaced by aplastic anemia, malignancies such as multiple myeloma, or infections such as tuberculosis, leading to a decrease in the production of blood cells and blood platelets. The bone marrow can also be affected by various forms of leukemia, which attacks its hematologic progenitor cells. Furthermore, exposure to radiation or chemotherapy will kill many of the rapidly dividing cells of the bone marrow, and will therefore result in a depressed immune system. Many of the symptoms of radiation poisoning are due to damage sustained by the bone marrow cells.

To diagnose diseases involving the bone marrow, a bone marrow aspiration is sometimes performed. This typically involves using a hollow needle to acquire a sample of red bone marrow from the crest of the ilium under general or local anesthesia.

Imaging

On CT and plain film, marrow change can be seen indirectly by assessing change to the adjacent ossified bone. Assessment with MRI is usually more sensitive and specific for pathology, particularly for hematologic malignancies like leukemia and lymphoma. These are difficult to distinguish from the red marrow hyperplasia of hematopoiesis, as can occur with tobacco smoking, chronically anemic disease states like sickle cell anemia or beta thalassemia, medications such as granulocyte colony-stimulating factors, or during recovery from chronic nutritional anemias or therapeutic bone marrow suppression. On MRI, the marrow signal is not supposed to be brighter than the adjacent intervertebral disc on T1 weighted images, either in the coronal or sagittal plane, where they can be assessed immediately adjacent to one another. Fatty marrow change, the inverse of red marrow hyperplasia, can occur with normal aging, though it can also be seen with certain treatments such as radiation therapy. Diffuse marrow T1 hypointensity without contrast enhancement or cortical discontinuity suggests red marrow conversion or myelofibrosis. Falsely normal marrow on T1 can be seen with diffuse multiple myeloma or leukemic infiltration when the water to fat ratio is not sufficiently altered, as may be seen with lower grade tumors or earlier in the disease process.

Histology

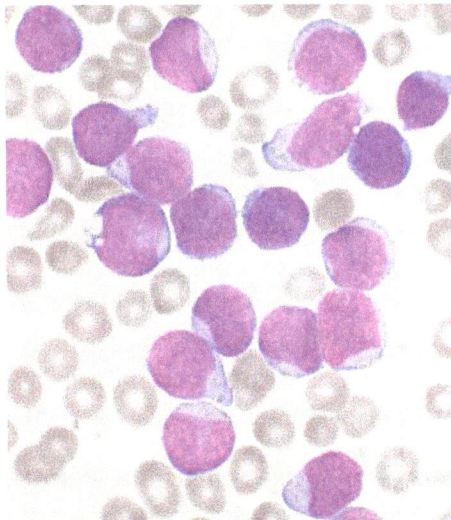

A Wright's-stained bone marrow aspirate smear from a patient with leukemia.

Bone marrow examination is the pathologic analysis of samples of bone marrow obtained via biopsy and bone marrow aspiration. Bone marrow examination is used in the diagnosis of a number of conditions, including leukemia, multiple myeloma, anemia, and pancytopenia. The bone marrow produces the cellular elements of the blood, including platelets, red blood cells and white blood cells. While much information can be gleaned by testing the blood itself (drawn from a vein by phlebotomy), it is sometimes necessary to examine the source of the blood cells in the bone marrow to obtain more information on hematopoiesis; this is the role of bone marrow aspiration and biopsy.

The ratio between myeloid series and erythroid cells is relevant to bone marrow function, and also to diseases of the bone marrow and peripheral blood, such as leukemia and anemia. The normal myeloid-to-erythroid ratio is around 3:1; this ratio may increase in myelogenous leukemias, decrease in polycythemias, and reverse in cases of thalassemia.

Donation and Transplantation

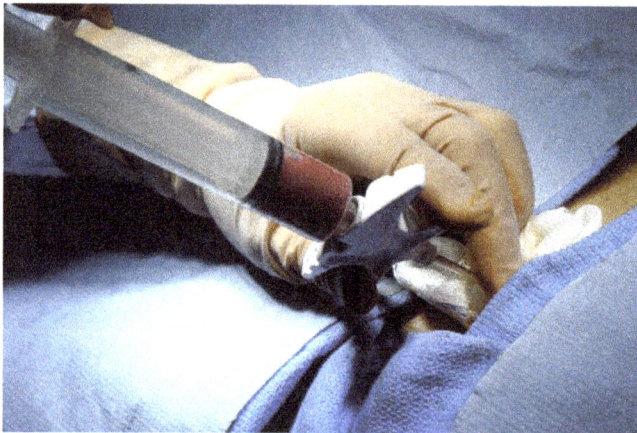

A bone marrow harvest in progress.

In a bone marrow transplant, hematopoietic stem cells are removed from a person and infused into another person (allogenic) or into the same person at a later time (autologous). If the donor and recipient are compatible, these infused cells will then travel to the bone marrow and initiate blood cell production. Transplantation from one person to another is conducted for the treatment of severe bone marrow diseases, such as congenital defects, autoimmune diseases or malignancies. The patient's own marrow is first killed off with drugs or radiation, and then the new stem cells are introduced. Before radiation therapy or chemotherapy in cases of cancer, some of the patient's hematopoietic stem cells are sometimes harvested and later infused back when the therapy is finished to restore the immune system.

Bone marrow stem cells can be induced to become neural cells to treat neurological illnesses, and can also potentially be used for the treatment of other illnesses, such as inflammatory bowel disease. In 2013, following a clinical trial, scientists proposed that bone marrow transplantation could be used to treat HIV in conjunction with antiretroviral drugs; however, it was later found that HIV remained in the bodies of the test subjects.

Harvesting

The stem cells are typically harvested directly from the red marrow in the iliac crest, often under general anesthesia. The procedure is minimally invasive and does not require stitches afterwards. Depending on the donor's health and reaction to the procedure, the actual harvesting can be an outpatient procedure, or can require 1–2 days of recovery in the hospital.

Another option is to administer certain drugs that stimulate the release of stem cells from the bone marrow into circulating blood. An intravenous catheter is inserted into the donor's arm, and the stem cells are then filtered out of the blood. This procedure is similar to that used in blood or platelet donation. In adults, bone marrow may also be taken from the sternum, while the tibia is often used when taking samples from infants. In newborns, stem cells may be retrieved from the umbilical cord.

Fossil Record

Bone marrow may have first evolved in *Eusthenopteron*, a species of prehistoric fish with close links to early tetrapods.

The earliest fossilised evidence of bone marrow was discovered in 2014 in *Eusthenopteron*, a lobe-finned fish which lived during the Devonian period approximately 370 million years ago. Scientists from Uppsala University and the European Synchrotron Radiation Facility used X-ray synchrotron microtomography to study the fossilised interior of the skeleton's humerus, finding organised tubular structures akin to modern vertebrate bone marrow. *Eusthenopteron* is closely related to the early tetrapods, which ultimately evolved into the land-dwelling mammals and lizards of the present day.

Endothelial Stem Cell

Endothelial stem cells (ESCs) are one of three types of stem cells found in bone marrow. They are multipotent, which describes the ability to give rise to many cell types, whereas a pluripotent stem cell can give rise to all types. ESCs have the characteristic properties of a stem cell: self-renewal and differentiation. These parent stem cells, ESCs, give rise to progenitor cells, which are intermediate stem cells that lose potency. Progenitor stem cells are committed to differentiating along a particular cell developmental pathway. ESCs will eventually produce endothelial cells (ECs), which create the thin-walled endothelium that lines the inner surface of blood vessels and lymphatic vessels.

Sources

ECs were first thought to arise from extraembryonic tissues because blood vessels were observed in the avian and mammalian embryos. However, after histological analysis, it was seen that ECs were in the embryo. This meant that blood vessels come from an intraembryonic source, the mesoderm.

Properties

Self-renewal and Differentiation

Stem cells have the unique ability make identical copies of themselves. This property maintains unspecialized and undifferentiated cells within the body. Differentiation is the process by which a cell becomes more specialized. For stem cells, this usually occurs through several stages, where a cell proliferates giving rise to daughter cells that are further specialized. For example, an endothelial progenitor cell (EPC) is more specialized than an ESC, and an EC is more specialized than an EPC. The further specialized a cell is, the more differentiated it is and as a result it is considered to be more committed to a certain cellular lineage.

Blood Vessel Formation

Blood vessels are made of a thin layer of ECs. As part of the circulatory system, blood vessels play a critical role in transporting blood throughout the body. Consequently, ECs have unique functions such as fluid filtration, homeostasis and hormone trafficking. ECs are the most differentiated form of an ESC. Formation of new blood vessels occurs by two different processes: vasculogenesis and angiogenesis. The former requires differentiation of endothelial cells from hemangioblasts and then the further organization into a primary capillary network. The latter occurs when new vessels are built from preexisting blood vessels.

Markers

The vascular system is made up of two parts: 1) Blood vasculature 2) Lymphatic vessels

Both parts consist of ECs that show differential expression of various genes. A study showed that ectopic expression of Prox-1 in blood vascular ECs (BECs) induced one-third of LEC specific gene expression. Prox-1is a homeobox transcription factor found in lymphatic ECs (LECs). For example, specific mRNAs such as VEGFR-3 and p57Kip2 were expressed by the BEC that was induced to express Prox-1.

Lymphatic-specific vascular endothelial growth factors VEGF-C and VEGF-D function as ligands for the vascular endothelial growth factor receptor 3 (VEGFR-3). The ligand-receptor interaction is essential for normal development of lymphatic tissues.

Tal1 gene is specifically found in the vascular endothelium and developing brain. This gene encodes the basic helix-loop-helix structure and functions as a transcription factor. Embryos lacking *Tal1* fail to develop past embryonic day 9.5. However, the study found that *Tal1* is actually required for vascular remodeling of the capillary network, rather than early endothelial development itself.

Fetal liver kinase-1 (Flk-1) is a cell surface receptor protein that is commonly used as a marker for ESCs and EPCs.

CD34 is another marker that can be found on the surface of ESCs and EPCs. It is characteristic of hematopoietic stem cells, as well as muscle stem cells.

Role in Formation of Vascular System

The two lineages arising from the EPC and the hematopoietic progenitor cell (HPC) form the blood circulatory system. Hematopoietic stem cells can of course undergo self-renewal, and are multipotent cells that give rise to erythrocytes (red blood cells), megakaryocytes/platelets, mast cells, T-lymphocytes, B-lymphocytes, dendritic cells, natural killer cells, monocyte/macrophage, and granulocytes. A study found that in the beginning stages of mouse embryogenesis, commencing at embryonic day 7.5, HPCs are produced close to the emerging vascular system. In the yolk sac's blood islands, HPCs and EC lineages emerge from the extraembryonic mesoderm in near unison. This creates a formation in which early erythrocytes are enveloped by angioblasts, and together they give rise to mature ECs. This observation gave rise to the hypothesis that the two lineages come from the same precursor, termed hemangioblast. Even though there is evidence that corroborates a hemangioblast, the isolation and exact location in the embryo has been difficult to pinpoint. Some researchers have found that cells with hemangioblast properties have been located in the posterior end of the primitive streak during gastrulation.

In 1917, Florence Sabin first observer of blood vessels and red blood cells in the yolk sac of chick embryos occur in close proximity and time. Then, in 1932, Murray detected the same event and created the term "hemangioblast" for what Sabin had seen.

Further evidence to corroborate hemangioblasts come from the expression of various genes such as CD34 and Tie2 by both lineages. The fact that this expression was seen in both EC and HPC lineages led researchers to propose a common origin. However, endothelial markers like Flk1/VEGFR-2 are exclusive to ECs but stop HPCs from progressing into an EC. It is accepted that VEGFR-2+ cells are a common precursor for HPCs and ECs. If the *Vegfr3* gene is deleted then both HPC and EC differentiation comes to a halt in embryos. VEGF promotes angioblast differentiation; whereas, VEGFR-1 stops the hemangioblast from becoming an EC. In addition, basic fibroblast growth factor FGF-2 is also involved in promoting angioblasts from the mesoderm. After angioblasts commit to becoming an EC, the angioblasts gather and rearrange to assemble in a tube similar to a capillary. Angioblasts can travel during the formation of the circulatory system to configure the branches to allow for directional blood flow. Pericytes and smooth muscle cells encircle ECs when they are differentiating into arterial or venous arrangements. Surrounding the ECs creates a brace to help stabilize the vessels known as the pericellular basal lamina. It is suggested pericytes and smooth muscle cells come from neural crest cells and the surrounding mesenchyme.

Role of Insulin-like Growth Factors in Endothelium Differentiation

ECs derived from stem cells are the beginning of vasculogenesis. Vasculogenesis is the new production of a vascular network from mesodermal progenitor cells. This can be distinguished from angiogenesis, which is the creation of new capillaries from vessels that already exist through the process of splitting or sprouting. This can occur "in vitro" in embryoid bodies (EB) derived from embryonic stem cells; this process in EB is similar to "in vivo" vasculogenesis. Important signaling factors for vasculogenesis are TGF-β, BMP4, and VEGF, all of which promote pluripotent stem cells to differentiate into mesoderm, endothelial progenitor cells, and then into mature endothelium.

It is well established that insulin-like growth factor (IGF) signaling is important for cell responses such as mitogenesis, cell growth, proliferation, angiogenesis, and differentiation. IGF1 and IGF2 increase the production of ECs in EB. A method that IGF employs to increase vasculogenesis is upregulation of VEGF. Not only is VEGF critical for mesoderm cells to become an EC, but also for EPCs to differentiate into mature endothelium. Understanding this process can lead to further research in vascular regeneration.

Animal Models of Vasculogenesis

There are a number of models used to study vasculogenesis. Avian embryos, Xenopus laevis embryos, are both fair models. However, zebrafish and mouse embryos have widespread use for easily observed development of vascular systems, and the recognition of key parts of molecular regulation when ECs differentiate.

Role in Recovery

ESCs and EPCs eventually differentiate into ECs. The endothelium secretes soluble factors to regulate vasodilatation and to preserve homeostasis. When there is any dysfunction in the endothelium, the body aims to repair the damage. Resident ESCs can generate mature ECs that replace the damaged ones. However, the intermediate progenitor cell cannot always generate functional ECs. This is because some of the differentiated cells may just have angiogenic properties.

Studies have shown that when vascular trauma occurs, EPCs and circulating endothelial progenitors (CEPs) are attracted to the site due to the release of specific chemokines. CEPs are derived from EPCs within the bone marrow, and the bone marrow is a reservoir of stem and progenitor cells. These cell types accelerate the healing process and prevent further complications such as hypoxia by gathering the cellular materials to reconstruct the endothelium.

Endothelium dysfunction is a prototypical characteristic of vascular disease, common in patients with autoimmune diseases such as systemic lupus erythematosus. Further, there is an inverse relationship between age and levels of EPCs. With a decline in EPCs the body loses its ability to repair the endothelium.

The use of stem cells for treatment has become a growing interest in the scientific community. Distinguishing between an ESC and its intermediate progenitor is nearly impossible, so research is now being done broadly on EPCs. One study showed that brief exposure to sevoflurane promoted growth and proliferation of EPCs. Sevoflurane is used in general anesthesia, but this finding shows the potential to induce endothelial progenitors. Using stem cells for cell replacement therapies is known as "regenerative medicine", which is a booming field that is now working on transplanting cells as opposed to bigger tissues or organs.

Role in Cancer

Understanding more about ESCs is important in cancer research. Tumours induce angiogenesis, which is the formation of new blood vessels. These cancerous cells do this by secreting factors such as VEGF and by reducing the amount of PGK, an anti-VEGF enzyme. The result is an uncontrolled production of beta-catenin, which regulates cell growth and cell mobility. With uncontrolled beta-catenin, the cell loses its adhesive properties. As ECs get packed together to create the lining for a new blood vessel, a single cancer cell is able to travel through the vessel to a distant site. If that cancer cell implants itself and begins forming a new tumour, the cancer has metastasized.

Future Efforts

Stem cells have always been a huge interest for scientists due to their unique properties that make them unlike any other cell in the body. Generally, the idea boils down to harnessing the power of plasticity and the ability to go from an unspecialized cell to a highly specialized differentiated cell. ESCs play an incredibly important role in establishing the vascular network that is vital for a functional circulatory system. Consequently, EPCs are under study to determine the potential for treatment of ischemic heart disease. Scientists are still trying to find a way to definitely distinguish the stem cell from the progenitor. In the case of endothelial cells, it is even difficult to distinguish a mature EC from an EPC. However, because of the multipotency of the ESC, the discoveries made about EPCs will parallel or understate the powers of the ESC.

Bone Marrow Examination

Bone marrow examination refers to the pathologic analysis of samples of bone marrow obtained by bone marrow biopsy (often called a trephine biopsy) and bone marrow aspiration. Bone marrow examination is used in the diagnosis of a number of conditions, including leukemia, multiple myeloma, lymphoma, anemia, and pancytopenia. The bone marrow produces the cellular elements of the blood, including platelets, red blood cells and white blood cells. While much information can be gleaned by testing the blood itself (drawn from a vein by phlebotomy), it is sometimes necessary to examine the source of the blood cells in the bone marrow to obtain more information on hematopoiesis; this is the role of bone marrow aspiration and biopsy.

Components of the Procedure

Section of bone marrow core biopsy as seen under the microscope (stained with H&E).

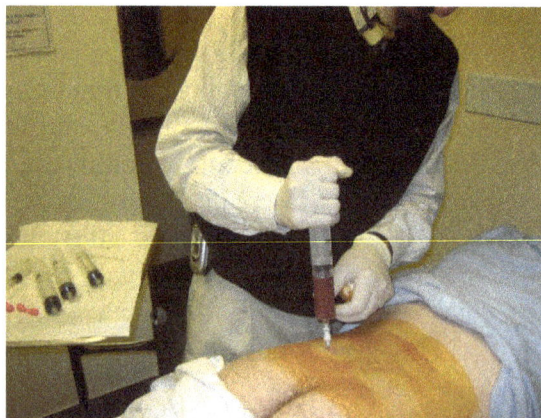

A volunteer donating bone marrow for scientific research.

Bone marrow samples can be obtained by aspiration and trephine biopsy. Sometimes, a bone marrow examination will include both an aspirate and a biopsy. The aspirate yields semi-liquid bone marrow, which can be examined by a pathologist under a light microscope and analyzed by flow cytometry, chromosome analysis, or polymerase chain reaction (PCR). Frequently, a trephine biopsy is also obtained, which yields a narrow, cylindrically shaped solid piece of bone marrow, 2mm wide and 2 cm long (80 µL), which is examined microscopically (sometimes with the aid of immunohistochemistry) for cellularity and infiltrative processes. An aspiration, using a 20 mL syringe, yields approximately 300 µL of bone marrow. A volume greater than 300 µL is not recommended, since it may dilute the sample with peripheral blood.

Comparison		
	Aspiration	**Biopsy**
Advantages	• Fast • Gives relative quantity of different cell types • Gives material to further study, e.g. molecular genetics and flow cytometry	• Gives cell and stroma constitution • Represents all cells • Explains cause of "dry tap" (aspiration gives no blood cells)
Drawbacks	Does not represent all cells	Slow processing

Aspiration does not always represent all cells since some such as lymphoma stick to the trabecula, and would thus be missed by a simple aspiration.

Site of Procedure

Bone marrow aspiration and trephine biopsy are usually performed on the back of the hipbone, or posterior iliac crest. An *aspirate* can also be obtained from the sternum (breastbone). For the sternal aspirate, the patient lies on their back, with a pillow under the shoulder to raise the chest. A *trephine biopsy* should never be performed on the sternum, due to the risk of injury to blood vessels, lungs or the heart. Bone marrow aspiration may also be performed on the tibial (shinbone) site in children up to 2 years of age while spinous process aspiration is frequently done in a lumbar puncture position and on the L3-L4 vertebrae.

Anesthesia is used to reduce surface pain at the spot where the needle is inserted. Pain may result from the procedure's insult to the marrow, which cannot be anesthetized, as well as short periods of pain from the anesthetic process itself. The experience is not uniform; different patients report different levels of pain, and some do not report any pain at certain expected points.

How the Test is Performed

A bone marrow biopsy may be done in a health care provider's office or in a hospital. Informed consent for the procedure is typically required. The patient is asked to lie on their abdomen (prone position) or on their side (lateral decubitus position). The skin is cleansed, and a local anesthetic such as lidocaine or procaine is injected to numb the area. Patients may also be pretreated with analgesics and/or anti-anxiety medications, although this is not a routine practice.

A needle used for bone marrow aspiration, with removable stylet.

Typically, the aspirate is performed first. An aspirate needle is inserted through the skin using manual pressure and force until it abuts the bone. Then, with a twisting motion of clinician's hand and wrist, the needle is advanced through the bony cortex (the hard outer layer of the bone) and into the marrow cavity. Once the needle is in the marrow cavity, a syringe is attached and used to aspirate ("suck out") liquid bone marrow. A twisting motion is performed during the aspiration to avoid excess content of blood in the sample, which might be the case if an excessively large sample from one single point is taken.

Subsequently, the biopsy is performed if indicated. A different, larger trephine needle is inserted and anchored in the bony cortex. The needle is then advanced with a twisting motion and rotated to obtain a solid piece of bone marrow. This piece is then removed along with the needle. The entire procedure, once preparation is complete, typically takes 10–15 minutes.

If several samples are taken, the needle is removed between the samples to avoid blood coagulation.

After the Procedure

After the procedure is complete, the patient is typically asked to lie flat for 5–10 minutes to provide pressure over the procedure site. After that, assuming no bleeding is observed, the patient can get up and go about their normal activities. Paracetamol (aka acetaminophen) or other simple analgesics can be used to ease soreness, which is common for 2–3 days after the procedure. Any worsening pain, redness, fever, bleeding or swelling may suggest a complication. Patients are also advised to avoid washing the procedure site for at least 24 hours after the procedure is completed.

Contraindications

There are few contraindications to bone marrow examination. It is important to note that thrombocytopenia or bleeding disorders are *not* contraindications as long as the procedure is performed by a skilled clinician. Bone marrow aspiration and biopsy can be safely performed even in the setting of extreme thrombocytopenia (low platelet count). If there is a skin or soft tissue infection over the hip, a different site should be chosen for bone marrow examination.

Complications

While mild soreness lasting 12–24 hours is common after a bone marrow examination, serious complications are extremely rare. In a large review, an estimated 55,000 bone marrow examinations were performed, with 26 serious adverse events (0.05%), including one fatality. The same author collected data on over 19,000 bone marrow examinations performed in the United Kingdom in 2003, and found 16 adverse events (0.08% of total procedures), the most common of which was bleeding. In this report, complications, while rare, were serious in individual cases.

References

- Raphael Rubin & David S. Strayer (2007). Rubin's Pathology: Clinicopathologic Foundations of Medicine. Lippincott Williams & Wilkins. p. 90. ISBN 0-7817-9516-8.

- Birbrair, Alexander; Frenette, Paul S. (2016-03-01). "Niche heterogeneity in the bone marrow". Annals of the New York Academy of Sciences: n/a–n/a. doi:10.1111/nyas.13016. ISSN 1749-6632.

- "The Red Cell Membrane: structure and pathologies" (PDF). Australian Centre for Blood Diseases/Monash University. Retrieved 24 January 2015.

- "Antibody Transforms Stem Cells Directly Into Brain Cells". Science Daily. 22 April 2013. Retrieved 24 April 2013.

- "Research Supports Promise of Cell Therapy for Bowel Disease". Wake Forest Baptist Medical Center. 28 February 2013. Retrieved 5 March 2013.

- "HIV Returns in Two Men Thought 'Cured' by Bone Marrow Transplants". RH Reality Check. 10 December 2013. Retrieved 10 December 2013.

- Bethesda MD. (6 April 2009). "Stem Cell Basics". In Stem Cell Information. National Institutes of Health, U.S. Department of Health and Human Services. Retrieved 6 March 2012.

- "Human acute myeloid leukemia is organized as a hierarchy that originates from a primitive hematopoietic cell". Nature. 1997. Retrieved 9 November 2012.

Significant Aspects of Stem Cell

The significant aspects of stem cell are cell potency, cellular differentiation, epigenetics in stem-cell differentiation and stem cell laws. Cell potency is the cell's capability of changing from one cell type to another whereas embryonic stem cells are cells, which can regenerate and differentiate as per the requirements within the body. The aspects elucidated are of vital importance, and provide a better understanding of stem cells.

Cell Potency

Cell potency is a cell's ability to differentiate into other cell types. The more cell types a cell can differentiate into, the greater its potency. Potency is also described as the gene activation potential within a cell which like a continuum begins with totipotency to designate a cell with the most differentiation potential, pluripotency, multipotency, oligopotency and finally unipotency. Potency is taken from the Latin term "potens" which means "having power."

Totipotency

Totipotency is the ability of a single cell to divide and produce all of the differentiated cells in an organism. Spores and zygotes are examples of totipotent cells. In the spectrum of cell potency, totipotency represents the cell with the greatest differentiation potential. *Toti* comes from the Latin *totus* which means "entirely".

It is possible for a fully differentiated cell to return to a state of totipotency. This conversion to totipotency is complex, not fully understood and the subject of recent research. Research in 2011 has shown that cells may differentiate not into a fully totipotent cell, but instead into a "complex cellular variation" of totipotency. Stem cells resembling totipotent blastomeres from 2-cell stage embryos can arise spontaneously in the embryonic stem cell cultures and also can be induced to arise more frequently in vitro through down-regulation of the chromatin assembly activity of CAF-1.

The human development model is one which can be used to describe how totipotent cells arise. Human development begins when a sperm fertilizes an egg and the resulting fertilized egg creates a single totipotent cell, a zygote. In the first hours after fertilization, this zygote divides into identical totipotent cells, which can later develop into any of the three germ layers of a human (endoderm, mesoderm, or ectoderm), into cells of the cytotrophoblast layer or syncytiotrophoblast layer of the placenta. After reaching a 16-cell stage, the totipotent cells of the morula differentiate into cells that will eventually become either the blastocyst's Inner cell mass or the outer trophoblasts. Approximately four days after fertilization and after several cycles of cell division, these totipotent cells begin to specialize. The inner cell mass, the source of embryonic stem cells, becomes pluripotent.

Research on *Caenorhabditis elegans* suggests that multiple mechanisms including RNA regulation may play a role in maintaining totipotency at different stages of development in some species. Work with zebrafish and mammals suggest a further interplay between miRNA and RNA binding proteins (RBPs) in determining development differences.

In September 2013, a team from the Spanish national Cancer Research Centre was able for the first time to make adult cells from mice retreat to the characteristics of embryonic stem cells, thereby achieving totipotency.

Pluripotency

In cell biology, pluripotency (from the Latin plurimus, meaning *very many*, and potens, meaning *having power*) refers to a stem cell that has the potential to differentiate into any of the three germ layers: endoderm (interior stomach lining, gastrointestinal tract, the lungs), mesoderm (muscle, bone, blood, urogenital), or ectoderm (epidermal tissues and nervous system). However, cell pluripotency is a continuum, ranging from the completely pluripotent cell that can form every cell of the embryo proper, e.g., embryonic stem cells and iPSCs to the incompletely or partially pluripotent cell that can form cells of all three germ layers but that may not exhibit all the charac-teristics of completely pluripotent cells.

Induced Pluripotency

Induced pluripotent stem cells, commonly abbreviated as iPS cells or iPSCs are a type of pluripotent stem cell artificially derived from a non-pluripotent cell, typically an adult somatic cell, by inducing a "forced" expression of certain genes and transcription factors. These transcription factors play a key role in determining the state of these cells and also highlights the fact that these somatic cells do preserve the same genetic information as early embryonic cells. The ability to induce cells into a pluripotent state was initially pioneered in 2006 using mouse fibroblasts and four transcription factors, Oct4, Sox2, Klf4 and c-Myc; this technique, called reprogramming, earned Shinya Yamanaka and John Gurdon the Nobel Prize in Physiology or Medicine 2012. This was then followed in 2007 by the successful induction of human iPSCs derived from human dermal fibroblasts using methods similar to those used for the induction of mouse cells. These induced cells exhibit similar traits to those of embryonic stem cells (ESCs) but do not require the use of embryos. Some of the similarities between ESCs and iPSCs include pluripotency, morphology, self-renewal ability, a trait that implies that they can divide and replicate indefinitely, and gene expression.

Epigenetic factors are also thought to be involved in the actual reprogramming of somatic cells in order to induce pluripotency. It has been theorized that certain epigenetic factors might actually work to clear the original somatic epigenetic marks in order to acquire the new epigenetic marks that are part of achieving a pluripotent state. Chromatin is also reorganized in iPSCs and becomes like that found in ESCs in that it is less condensed and therefore more accessible. Euchromatin modifications are also common which is also consistent with the state of euchromatin found in ESCs.

Due to their great similarity to ESCs, iPSCs have been of great interest to the medical and research community. iPSCs could potentially have the same therapeutic implications and applications as ESCs but without the controversial use of embryos in the process, a topic of great bioethical debate. In fact, the induced pluripotency of somatic cells into undifferentiated iPS cells was originally hailed as the end of the controversial use of embryonic stem cells. However, iPSCs were found to be potentially tumorigenic, and, despite advances, were never approved for clinical stage research in the United States. Setbacks such as low replication rates and early senescence have also been encountered when making iPSCs, hindering their use as ESCs replacements.

Additionally, it has been determined that the somatic expression of combined transcription factors can directly induce other defined somatic cell fates (transdifferentiation); researchers identified three neural-lineage-specific transcription factors that could directly convert mouse fibroblasts (skin cells) into fully functional neurons. This result challenges the terminal nature of cellular differentiation and the integrity of lineage commitment; and implies that with the proper tools, *all* cells are totipotent and may form all kinds of tissue.

Some of the possible medical and therapeutic uses for iPSCs derived from patients include their use in cell and tissue transplants without the risk of rejection that is commonly encountered. iPSCs can potentially replace animal models unsuitable as well as in-vitro models used for disease research.

Multipotency

Multipotency describes progenitor cells which have the gene activation potential to differentiate into multiple, but limited cell types. For example, a multipotent blood stem cell is a hematopoietic cell — and this cell type can differentiate itself into several types of blood cell types like lymphocytes, monocytes, neutrophils, etc., but cannot differentiate into brain cells, bone cells or other non-blood cell types.

New research related to multipotent cells suggests that multipotent cells may be capable of conversion into unrelated cell types. In one case, fibroblasts were converted into functional neurons. In another case, human umbilical cord blood stem cells were converted into human neurons. Research is also focusing on converting multipotent cells into pluripotent cells.

Multipotent cells are found in many, but not all human cell types. Multipotent cells have been found in cord blood, adipose tissue, cardiac cells, bone marrow, and mesenchymal stem cells (MSCs) which are found in the third molar.

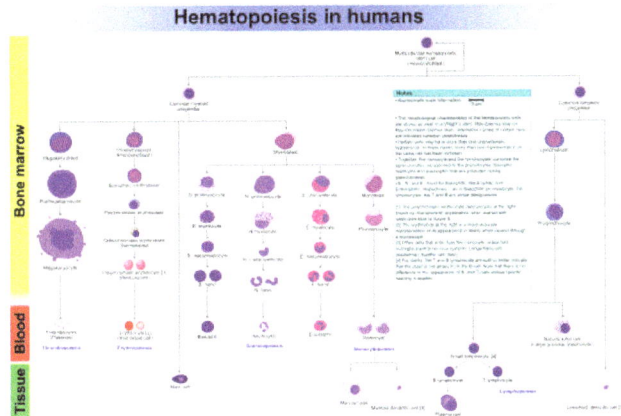

Hematopoietic stem cells are an example of multipotency. When they differentiate into myeloid or lymphoid progenitor cells, they lose potency and become oligopotent cells with the ability to give rise to all cells of its lineage.

MSCs may prove to be a good, reliable source for stem cells because of the ease in collection of molars at 8–10 years of age and before adult dental calcification. MSCs can differentiate into osteoblasts, chondrocytes, and adipocytes.

Oligopotency

In biology, oligopotency is the ability of progenitor cells to differentiate into a few cell types. It is a degree of potency. Examples of oligopotent stem cells are the lymphoid or myeloid stem cells. A lymphoid cell specifically, can give rise to various blood cells such as B and T cells, however, not to a different blood cell type like a red blood cell. Examples of progenitor cells are vascular stem cells that have the capacity to become both endothelial or smooth muscle cells.

Unipotency

In cell biology, a unipotent cell is the concept that one stem cell has the capacity to differentiate into only one cell type. It is currently unclear if true unipotent stem cells exist. Hepatoblasts, which differentiate into hepatocytes (which constitute most of the liver) or cholangiocytes (epithelial cells of the bile duct), are bipotent. A close synonym for *unipotent cell* is *precursor cell*.

Cellular Differentiation

In developmental biology, cellular differentiation is the process where a cell changes from one cell type to another. Most commonly this is a less specialized type becoming a more specialized type, such as during cell growth. Differentiation occurs numerous times during the development of a multicellular organism as it changes from a simple zygote to a complex system of tissues and cell types. Differentiation continues in adulthood as adult stem cells divide and create fully differentiated daughter cells during tissue repair and during normal cell turnover. Some differentiation occurs in response to antigen exposure. Differentiation dramatically changes a cell's size, shape, membrane potential, metabolic activity, and responsiveness to signals. These changes are largely due to highly controlled modifications in gene expression and are the study of epigenetics. With a few exceptions, cellular differentiation almost never involves a change in the DNA sequence itself. Thus, different cells can have very different physical characteristics despite having the same genome.

A cell that can differentiate into all cell types of the adult organism is known as *pluripotent*. Such cells are called embryonic stem cells in animals and meristematic cells in higher plants. A cell that can differentiate into all cell types, including the placental tissue, is known as *totipotent*. In mammals, only the zygote and subsequent blastomeres are totipotent, while in plants many differentiated cells can become totipotent with simple laboratory techniques. In cytopathology, the level of cellular differentiation is used as a measure of cancer progression. "Grade" is a marker of how differentiated a cell in a tumor is.

Mammalian Cell Types

Three basic categories of cells make up the mammalian body: germ cells, somatic cells, and stem cells. Each of the approximately 100 trillion (10^{14}) cells in an adult human has its own copy or copies of the genome except certain cell types, such as red blood cells, that lack nuclei in their fully differentiated state. Most cells are diploid; they have two copies of each chromosome. Such cells, called somatic cells, make up most of the human body, such as skin and muscle cells. Cells differentiate to specialize for different functions.

Germ line cells are any line of cells that give rise to gametes—eggs and sperm—and thus are continuous through the generations. Stem cells, on the other hand, have the ability to divide for indefinite periods and to give rise to specialized cells. They are best described in the context of normal human development. Development begins when a sperm fertilizes an egg and creates a single cell that has the potential to form an entire organism. In the first hours after fertilization, this cell divides into identical cells. In humans, approximately four days after fertilization and after several cycles of cell division, these cells begin to specialize, forming a hollow sphere of cells, called a blastocyst. The blastocyst has an outer layer of cells, and inside this hollow sphere, there is a cluster of cells called the inner cell mass. The cells of the inner cell mass go on to form virtually all of the tissues of the human body. Although the cells of the inner cell mass can form virtually every type of cell found in the human body, they cannot form an organism. These cells are referred to as pluripotent.

Pluripotent stem cells undergo further specialization into multipotent progenitor cells that then give rise to functional cells. Examples of stem and progenitor cells include:*Radial glial cells* (embryonic neural stem cells) that give rise to excitatory neurons in the fetal brain through the process of neurogenesis.

- *Hematopoietic stem cells* (adult stem cells) from the bone marrow that give rise to red blood cells, white blood cells, and platelets

- *Mesenchymal stem cells* (adult stem cells) from the bone marrow that give rise to stromal cells, fat cells, and types of bone cells

- *Epithelial stem cells* (progenitor cells) that give rise to the various types of skin cells

- *Muscle satellite cells* (progenitor cells) that contribute to differentiated muscle tissue.

A pathway that is guided by the cell adhesion molecules consisting of four amino acids, arginine, glycine, asparagine, and serine, is created as the cellular blastomere differentiates from the single-layered blastula to the three primary layers of germ cells in mammals, namely the ectoderm, mesoderm and endoderm (listed from most distal (exterior) to proximal (interior)). The ectoderm ends up forming the skin and the nervous system, the mesoderm forms the bones and muscular tissue, and the endoderm forms the internal organ tissues.

Dedifferentiation

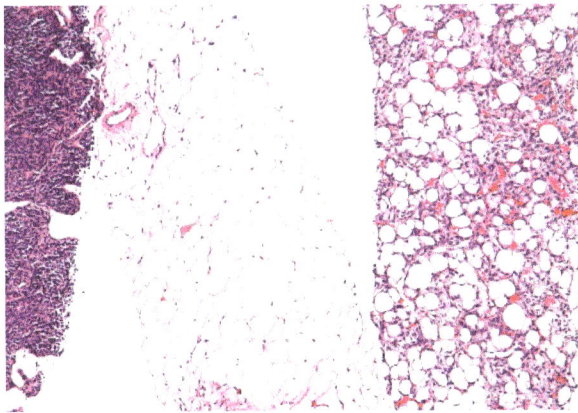

Micrograph of a liposarcoma with some dedifferentiation, that is not identifiable as a liposarcoma, (left edge of image) and a differentiated component (with lipoblasts and increased vascularity (right of image)). Fully differentiated (morphologically benign) adipose tissue (center of the image) has few blood vessels. H&E stain.

Dedifferentiation, or integration is a cellular process often seen in more basal life forms such as worms and amphibians in which a partially or terminally differentiated cell reverts to an earlier developmental stage, usually as part of a regenerative process.

Dedifferentiation also occurs in plants. Cells in cell culture can lose properties they originally had, such as protein expression, or change shape. This process is also termed dedifferentiation.

Some believe dedifferentiation is an aberration of the normal development cycle that results in cancer, whereas others believe it to be a natural part of the immune response lost by humans at some point as a result of evolution.

A small molecule dubbed reversine, a purine analog, has been discovered that has proven to induce dedifferentiation in myotubes. These dedifferentiated cells could then re-differentiate into osteoblasts and adipocytes.

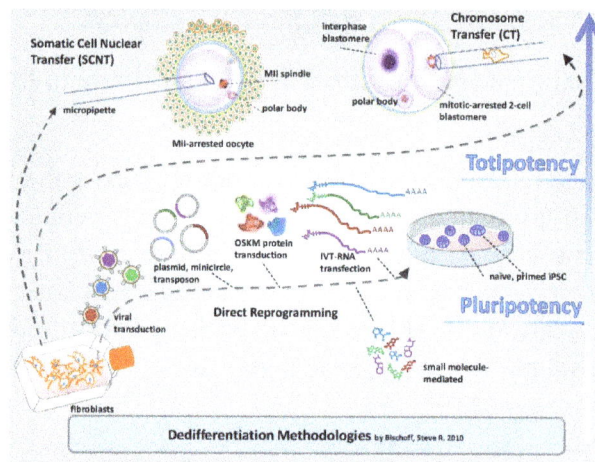

Diagram exposing several methods used to revert adult somatic cells to totipotency or pluripotency.

Mechanisms

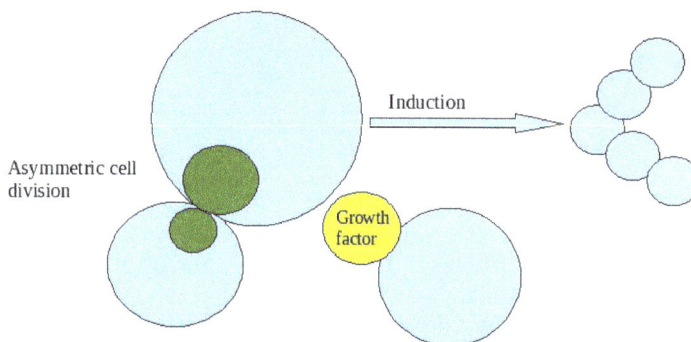

Mechanisms of cellular differentiation.

Each specialized cell type in an organism expresses a subset of all the genes that constitute the genome of that species. Each cell type is defined by its particular pattern of regulated gene expression. Cell differentiation is thus a transition of a cell from one cell type to another and it involves a switch from one pattern of gene expression to another.

Cellular differentiation during development can be understood as the result of a gene regulatory network. A regulatory gene and its cis-regulatory modules are nodes in a gene regulatory network; they receive input and create output elsewhere in the network. The systems biology approach to developmental biology emphasizes the importance of investigating how developmental mechanisms interact to produce predictable patterns (morphogenesis). (However, an alternative view has been proposed recently. Based on stochastic gene expression, cellular differentiation is the result of a Darwinian selective process occurring among cells. In this frame, protein and gene networks are the result of cellular processes and not their cause. See: Cellular Darwinism)

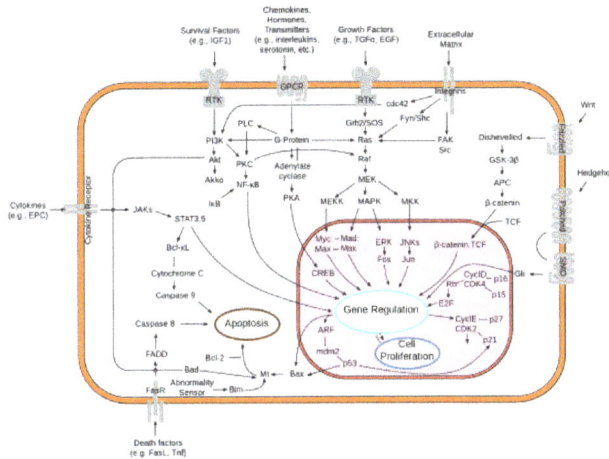

An overview of major signal transduction pathways.

A few evolutionarily conserved types of molecular processes are often involved in the cellular mechanisms that control these switches. The major types of molecular processes that control cellular differentiation involve cell signaling. Many of the signal molecules that convey information from cell to cell during the control of cellular differentiation are called growth factors. Although the details of specific signal transduction pathways vary, these pathways often share the following general steps. A ligand produced by one cell binds to a receptor in the extracellular region of another cell, inducing a conformational change in the receptor. The shape of the cytoplasmic domain of the receptor changes, and the receptor acquires enzymatic activity. The receptor then catalyzes reactions that phosphorylate other proteins, activating them. A cascade of phosphorylation reactions eventually activates a dormant transcription factor or cytoskeletal protein, thus contributing to the differentiation process in the target cell. Cells and tissues can vary in competence, their ability to respond to external signals.

Signal induction refers to cascades of signaling events, during which a cell or tissue signals to another cell or tissue to influence its developmental fate. Yamamoto and Jeffery investigated the role of the lens in eye formation in cave- and surface-dwelling fish, a striking example of induction. Through reciprocal transplants, Yamamoto and Jeffery found that the lens vesicle of surface fish can induce other parts of the eye to develop in cave- and surface-dwelling fish, while the lens vesicle of the cave-dwelling fish cannot.

Other important mechanisms fall under the category of asymmetric cell divisions, divisions that give rise to daughter cells with distinct developmental fates. Asymmetric cell divisions can occur because of asymmetrically expressed maternal cytoplasmic determinants or because of signaling. In the former mechanism, distinct daughter cells are created during cytokinesis because of an uneven distribution of regulatory molecules in the parent cell; the distinct cytoplasm that each daughter cell inherits results in a distinct pattern of differentiation for each daughter cell. A well-studied example of pattern formation by asymmetric divisions is body axis patterning in Drosophila. RNA molecules are an important type of intracellular differentiation control signal. The molecular and genetic basis of asymmetric cell divisions has also been studied in green algae of the genus *Volvox*, a model system for studying how unicellular organisms can evolve into multicellular organisms. In *Volvox carteri*, the 16 cells in the anterior hemisphere of a 32-cell embryo divide asymmetrically, each producing one large and one small daughter cell. The size of the cell at the end of all cell divisions determines whether it becomes a specialized germ or somatic cell.

Epigenetic Control of Cellular Differentiation

Since each cell, regardless of cell type, possesses the same genome, determination of cell type must occur at the level of gene expression. While the regulation of gene expression can occur through cis- and trans-regulatory elements including a gene's promoter and enhancers, the problem arises as to how this expression pattern is maintained over numerous generations of cell division. As it turns out, epigenetic processes play a crucial role in regulating the decision to adopt a stem, progenitor, or mature cell fate. This section will focus primarily on mammalian stem cells.

In systems biology and mathematical modeling of gene regulatory networks, cell-fate determination is predicted to exhibit certain dynamics, such as attractor-convergence (the attractor can be an equilibrium point, limit cycle or strange attractor) or oscillatory.

Importance of Epigenetic Control

The first question that can be asked is the extent and complexity of the role of epigenetic processes in the determination of cell fate. A clear answer to this question can be seen in the 2011 paper by Lister R, *et al.* on aberrant epigenomic programming in human induced pluripotent stem cells. As induced pluripotent stem cells (iPSCs) are thought to mimic embryonic stem cells in their pluripotent properties, few epigenetic differences should exist between them. To test this prediction, the authors conducted whole-genome profiling of DNA methylation patterns in several human embryonic stem cell (ESC), iPSC, and progenitor cell lines.

Female adipose cells, lung fibroblasts, and foreskin fibroblasts were reprogrammed into induced pluripotent state with the OCT4, SOX2, KLF4, and MYC genes. Patterns of DNA methylation in ESCs, iPSCs, somatic cells were compared. Lister R, *et al.* observed significant resemblance in methylation levels between embryonic and induced pluripotent cells. Around 80% of CG dinucleotides in ESCs and iPSCs were methylated, the same was true of only 60% of CG dinucleotides in somatic cells. In addition, somatic cells possessed minimal levels of cytosine methylation in non-CG dinucleotides, while induced pluripotent cells possessed similar levels of methylation as embryonic stem cells, between 0.5 and 1.5%. Thus, consistent with their respective transcriptional activities, DNA methylation patterns, at least on the genomic level, are similar between ESCs and iPSCs.

However, upon examining methylation patterns more closely, the authors discovered 1175 regions of differential CG dinucleotide methylation between at least one ES or iPS cell line. By comparing these regions of differential methylation with regions of cytosine methylation in the original somatic cells, 44-49% of differentially methylated regions reflected methylation patterns of the respective progenitor somatic cells, while 51-56% of these regions were dissimilar to both the progenitor and embryonic cell lines. In vitro-induced differentiation of iPSC lines saw transmission of 88% and 46% of hyper and hypo-methylated differentially methylated regions, respectively.

Two conclusions are readily apparent from this study. First, epigenetic processes are heavily involved in cell fate determination, as seen from the similar levels of cytosine methylation between induced pluripotent and embryonic stem cells, consistent with their respective patterns of transcription. Second, the mechanisms of de-differentiation (and by extension, differentiation) are very complex and cannot be easily duplicated, as seen by the significant number of differentially methylated regions between ES and iPS cell lines. Now that these two points have been established, we can examine some of the epigenetic mechanisms that are thought to regulate cellular differentiation.

Mechanisms of Epigenetic Regulation

Pioneering Factors (Oct4, Sox2, Nanog)

Three transcription factors, OCT4, SOX2, and NANOG – the first two of which are used in iPSC reprogramming – are highly expressed in undifferentiated embryonic stem cells and are necessary for the maintenance of their pluripotency. It is thought that they achieve this through alterations in chromatin structure, such as histone modification and DNA methylation, to restrict or permit the transcription of target genes.

Polycomb Repressive Complex (PRC2)

In the realm of gene silencing, Polycomb repressive complex 2, one of two classes of the Polycomb group (PcG) family of proteins, catalyzes the di- and tri-methylation of histone H3 lysine 27 (H3K27me2/me3). By binding to the H3K27me2/3-tagged nucleosome, PRC1 (also a complex of PcG family proteins) catalyzes the mono-ubiquitinylation of histone H2A at lysine 119 (H2AK119Ub1), blocking RNA polymerase II activity and resulting in transcriptional suppression. PcG knockout ES cells do not differentiate efficiently into the three germ layers, and deletion of the PRC1 and PRC2 genes leads to increased expression of lineage-affiliated genes and unscheduled differentiation. Presumably, PcG complexes are responsible for transcriptionally repressing differentiation and development-promoting genes.

Trithorax Group Proteins (TrxG)

Alternately, upon receiving differentiation signals, PcG proteins are recruited to promoters of pluripotency transcription factors. PcG-deficient ES cells can begin differentiation but cannot maintain the differentiated phenotype. Simultaneously, differentiation and development-promoting genes are activated by Trithorax group (TrxG) chromatin regulators and lose their repression. TrxG proteins are recruited at regions of high transcriptional activity, where they catalyze the trimethylation of histone H3 lysine 4 (H3K4me3) and promote gene activation through histone acetylation. PcG and TrxG complexes engage in direct competition and are thought to be functionally antagonistic, creating at differentiation and development-promoting loci what is termed a "bivalent domain" and rendering these genes sensitive to rapid induction or repression.

DNA Methylation

Regulation of gene expression is further achieved through DNA methylation, in which the DNA methyltransferase-mediated methylation of cytosine residues in CpG dinucleotides maintains heritable repression by controlling DNA accessibility. The majority of CpG sites in embryonic stem cells are unmethylated and appear to be associated with H3K4me3-carrying nucleosomes. Upon differentiation, a small number of genes, including OCT4 and NANOG, are methylated and their promoters repressed to prevent their further expression. Consistently, DNA methylation-deficient embryonic stem cells rapidly enter apoptosis upon in vitro differentiation.

Nucleosome Positioning

While the DNA sequence of most cells of an organism is the same, the binding patterns of transcription factors and the corresponding gene expression patterns are different. To a large extent, differences in transcription factor binding are determined by the chromatin accessibility of their binding sites through histone modification and/or pioneer factors. In particular, it is important to know whether a nucleosome is covering a given genomic binding site or not. Recent studies have elucidated the role of nucleosome positioning during stem cell development.

Role of Signaling in Epigenetic Control

A final question to ask concerns the role of cell signaling in influencing the epigenetic processes governing differentiation. Such a role should exist, as it would be reasonable to think that extrinsic signaling can lead to epigenetic remodeling, just as it can lead to changes in gene expression through the activation or repression of different transcription factors. Interestingly, little direct data is available concerning the specific signals that influence the epigenome, and the majority of current knowledge consist of speculations on plausible candidate regulators of epigenetic remodeling. We will first discuss several major candidates thought to be involved in the induction and maintenance of both embryonic stem cells and their differentiated progeny, and then turn to one example of specific signaling pathways in which more direct evidence exists for its role in epigenetic change.

The first major candidate is Wnt signaling pathway. The Wnt pathway is involved in all stages of differentiation, and the ligand Wnt3a can substitute for the overexpression of c-Myc in the generation of induced pluripotent stem cells. On the other hand, disruption of ß-catenin, a component of the Wnt signaling pathway, leads to decreased proliferation of neural progenitors.

Growth factors comprise the second major set of candidates of epigenetic regulators of cellular differentiation. These morphogens are crucial for development, and include bone morphogenetic proteins, transforming growth factors (TGFs), and fibroblast growth factors (FGFs). TGFs and FGFs have been shown to sustain expression of OCT4, SOX2, and NANOG by downstream signaling to Smad proteins. Depletion of growth factors promotes the differentiation of ESCs, while genes with bivalent chromatin can become either more restrictive or permissive in their transcription.

Several other signaling pathways are also considered to be primary candidates. Cytokine leukemia inhibitory factors are associated with the maintenance of mouse ESCs in an undifferentiated state. This is achieved through its activation of the Jak-STAT3 pathway, which has been shown to be necessary and sufficient towards maintaining mouse ESC pluripotency. Retinoic acid can induce differentiation of human and mouse ESCs, and Notch signaling is involved in the proliferation and self-renewal of stem cells. Finally, Sonic hedgehog, in addition to its role as a morphogen, promotes embryonic stem cell differentiation and the self-renewal of somatic stem cells.

The problem, of course, is that the candidacy of these signaling pathways was inferred primarily on the basis of their role in development and cellular differentiation. While epigenetic regulation is necessary for driving cellular differentiation, they are certainly not sufficient for this process. Direct modulation of gene expression through modification of transcription factors plays a key role that must be distinguished from heritable epigenetic changes that can persist even in the absence of the original environmental signals. Only a few examples of signaling pathways leading to epigenetic changes that alter cell fate currently exist, and we will focus on one of them.

Expression of Shh (Sonic hedgehog) upregulates the production of BMI1, a component of the PcG complex that recognizes H3K27me3. This occurs in a Gli-dependent manner, as Gli1 and Gli2 are downstream effectors of the Hedgehog signaling pathway. In culture, Bmi1 mediates the Hedgehog pathway's ability to promote human mammary stem cell self-renewal. In both humans and mice, researchers showed Bmi1 to be highly expressed in proliferating immature cerebellar granule cell precursors. When Bmi1 was knocked out in mice, impaired cerebellar development resulted, leading to significant reductions in postnatal brain mass along with abnormalities in motor control and behavior. A separate study showed a significant decrease in neural stem cell proliferation along with increased astrocyte proliferation in Bmi null mice.

In summary, the role of signaling in the epigenetic control of cell fate in mammals is largely unknown, but distinct examples exist that indicate the likely existence of further such mechanisms.

Effect of Matrix Elasticity on Differentiation

In order to fulfill the purpose of regenerating a variety of tissues, adult stems are known to migrate from their niches, adhere to new extracellular matrices (ECM) and differentiate. The ductility of these microenvironments are unique to different tissue types. The ECM surrounding brain, muscle and bone tissues range from soft to stiff. The transduction of the stem cells into these cells types is not directed solely by chemokine cues and cell to cell signaling. The elasticity of the microenvironment can also affect the differentiation of mesenchymal stem cells (MSCs which originate in bone marrow.) When MSCs are placed on substrates of the same stiffness as brain, muscle and bone ECM, the MSCs take on properties of those respective cell types. Matrix sensing requires the cell to pull against the matrix at focal adhesions, which triggers a cellular mechano-transducer to generate a signal to be informed what force is needed to deform the matrix. To determine the key players in matrix-elasticity-driven lineage specification in MSCs, different matrix microenvironments were mimicked. From these experiments, it was concluded that focal adhesions of the MSCs were the cellular mechano-transducer sensing the differences of the matrix elasticity. The non-muscle myosin IIa-c isoforms generates the forces in the cell that lead to signaling of early commitment markers. Nonmuslce myosin IIa generates the least force increasing to non-muscle myosin IIc. There are also factors in the cell that inhibit non-muscle myosin II, such as blebbistatin. This makes the cell effectively blind to the surrounding matrix. Researchers have obtained some success in inducing stem cell-like properties in HEK 239 cells by providing a soft matrix without the use of diffusing factors. The stem-cell properties appear to be linked to tension in the cells' actin network. One identified mechanism for matrix-induced differentiation is tension-induced proteins, which remodel chromatin in response to mechanical stretch. The RhoA pathway is also implicated in this process.

Epigenetics in Stem-cell Differentiation

Embryonic stem cells are capable of self-renewing and differentiating to the desired fate depending on its position within the body. Stem cell homeostasis is maintained through epigenetic mechanisms that are highly dynamic in regulating the chromatin structure as well as specific gene transcription programs. Epigenetics has been used to refer to changes in gene expression, which are heritable through modifications not affecting the DNA sequence.

The mammalian epigenome undergoes global remodeling during early stem cell development that requires commitment of cells to be restricted to the desired lineage. There has been multiple evidence suggesting that the maintenance of the lineage commitment of stem cells are controlled by epigenetic mechanisms such as DNA methylation, histone modifications and regulation of ATP-dependent remolding of chromatin structure. Based on the *histone code* hypothesis, distinct covalent histone modifications can lead to functionally distinct chromatin structures that influence the fate of the cell.

This regulation of chromatin through epigenetic modifications is a molecular mechanism that will determine whether the cell will continue to differentiate into the desired fate. A research study performed by *Lee et al.* examined the effects of epigenetic modifications on the chromatin structure and the modulation of these epigenetic markers during stem cell differentiation through in vitro differentiation of murine embryonic stem (ES) cells.

Experimental Background

Embryonic stem cells exhibit dramatic and complex alterations to both global and site-specific chromatin structures. *Lee et al.* performed an experiment to determine the importance of deacetylation and acetylation for stem cell differentiation by looking at global acetylation and methylation levels at certain site-specific modification in histone sites $H3K9$ and $H3K4$. Gene expression at these histones regulated by epigenetic modifications is critical in restricting the embryonic stem cell to desired cell lineages and developing cellular memory.

For mammalian cells, the maintenance of cytosine methylation is catalyzed by DNA methyltransferases and any disruption to these methyltransferases will cause a lethal phenotype to the embryo. Cytosine methylation is examined at $H3K9$, which is associated with inactive heterochromatin and occurs mainly at CpG dinucleotides while global acetylation is examined at $H3K4$, which is associated with active euchromatin. The mammalian zygotic genome undergoes active and passive global cytosine demethylation following fertilization that reaches a minimal point of 20% CpG methylation at the blastocyst stage to which is then followed by a wave of methylation that reprograms the chromatin structure in order to restore global levels of CpG methylation to 60%. Embryonic stem cells containing reduced or elevation levels of methylation are viable but unable to differentiate and therefore require critical regulation of cytosine methylation for mammalian development.

Effects of Global Histone Modifications During Embryonic Stem Cell Differentiation

Histones modifications in chromatin were analyzed at various time intervals (along a

6 day period) following the initiation of in vitro embryonic stem cell differentiation. Differentiation was triggered by the removal of Leukemia inhibitory factor (LIF) which inhibits differentiation. Representative data of the histone modifications at the specific sites were assessed using Western blotting. The data confirms that strong deacetylation at the *H3K4* and *H3K9* positions of histone H3 one day after LIF removal, followed by a small increase in acetylation by day two.

The histone *H3K4* methylation also decreased after one day of LIF removal but showed a rebound between days 2-4 of differentiation, finally ending with a decrease in methylation on day five. These results indicate a decrease in the level of active euchromatin epigenetic marks upon initiation of embryonic stem cell differentiation which is then followed immediately by reprogramming of the epigenome.

Histone modifications of *H3K9* position show a decrease in di- and tri-methylation of undifferentiated embryonic stem cells and had a gradual increase in methylation during the six-day time course of in vitro differentiation, which indicated that there is a global increase of inactive heterochromatin levels at this histone mark.

As the embryonic stem cell undergoes differentiation the markers for active euchromatin (histone acetylation and *H3K4* methylation) are decreased after the removal of LIF showing that the cell is indeed becoming more differentiated. The slight rebound in each of these marks allows for further differentiation to occur by allowing another opportunity to decrease the markers once again, bringing the cell closer to its desired fate. Since there is also an increase throughout the six-day period in *H3K9me*, a marker for active heterochromatin, once differentiation occurs it is concluded that the formation of heterochromatin occurs as the cell is differentiated into its desired fate making the cell inactive to prevent further differentiation.

DNA Methylation in Differentiated Versus Undifferentiated Cells

Global levels of cytosine methylation were compared between undifferentiated and differentiated embryonic stem cells. Global 5-methylcytosine levels have been measured prior to differentiation and after in vitro differentiation. The global cytosine methylation pattern appears to be established prior to the reprogramming of the histone code that occurs upon in vitro differentiation of embryonic stem cells.

As the embryonic stem cell undergoes differentiation the level of DNA methylation increases. This is in agreement with findings that show that there is an increase in inactive heterochromatin during differentiation.

Supplemental Effects of Methylation with DNMTs

In mammals, DNA methylation plays a role in regulating a key component of multipotency—the ability to rapidly self-renew. Khavari et al. discussed the fundamental mechanisms of DNA methylation and the interaction with several pathways regulating differentiation. New approaches studying the genomic status of DNA methylation in various states of differentiation have shown that methylation at CpG sites associated with putative enhancers are important in this process. DNA methylation can modulate the binding affinities of transcription factors by recruiting repressors such as *MeCP2* which display binding specificity for sequences containing methylated CpG dinucleotides. DNA methylation is controlled by certain methyltransferases, *DMNTs*, which perform different functions depending on each one. *DNMT3A* and *DNMT3B* have both been linked to a role in the establishment of DNA methylation pattern in the early development of the stem cell, whereas *DNMT1* is required to methylate a newly synthesized strand of DNA after the cell has undergone replication in order to sustain the epigenetic regulatory state. Numerous proteins can physically interact with *DNMTs* themselves, which help target *DNMT1* to hemi-methylated DNA.

Several new studies point to the central role of DNA methylation interacting with the regulation of cell cycles and DNA repair pathways in order to maintain the undifferentiated state. In embryonic stem cells, *DNMT1* depletion within the undifferentiated progenitor cell compartment led to cell cycle arrest, premature differentiation and a failure of tissue self-renewal. The loss of *DNMT1* occurred from profound effects associated with activation of differentiation genes and loss of genes promoting cell cycle progression, thus indicating that *DNMT1* and other *DNMTs* do not continuously suppress differentiation and thus maintain the pluripotent state.

These studies point to the important of the interaction of DNMTs in order to maintain stem cell states allowing for further differentiation and formation of heterochromatin to occur.

Epigenetic Modifications of Regulated Genes During ESC Differentiation

Okamoto et al. previously documented the expression level of the *Oct4* gene decreasing with embryonic stem cell differentiation. *Lee et al* performed a ChIP analysis of the Oct4 promoter, associated with undifferentiated cells, region to examine the epigenetic modifications of regulated genes undergoing development during embryonic stem cell differentiation. This promoter region decreased at *H3K4* methylation and *H3K9* acetylation sites and increased at the *H3K9* methylation site during differentiation. Analysis of a CpG motif of the *Oct4* gene promoter revealed a progressive increase of DNA methylation and was completely methylated at day 10 of differentiation as previously reported in Gidekel and Bergman.

These results indicate that there was a shift from the active eurchromatin to the inactive heterochromatin due to the decrease of acetylation of H3K4 and an increase of H3K9me. This means that the cell is becoming differentiated at the Oct4 gene, which is coincident with the silence of Oct4 gene expression.

Another site specific gene tested for histone modification was a *Brachyury* gene, a marker of mesoderm differentiation and is only slightly expressed in undifferentiated embryonic stem cells. "Brachyury" was induced at day five of differentiation and completely silencing by day 10, corresponding to the last day of differentiation. The ChIP analysis of the "Brachyury" gene promoter revealed increase of expression in mono- and di-methylation of $H3K4$ at day 0 and 5 of embryonic stem cell differentiation with a loss of gene expression at day 10. $H3K4$ trimethylation coincides with the time of highest Brachyury gene expression since it only had gene expression on day 5. $H3K4$ methylations in all forms are absent at day 10 of differentiation, which correlates with the silencing of Brachyury gene expression. Mono-methylation of both histones produced expression at day 0 indication a marker that is not useful for chromatin structure. Acetlyation of H3K9 does not correlate to *Brachyury* gene expression since it was down regulated at the induction of differentiation. Upon examining of DNA methylation expression, there was no formation of intermediate sized bad in the Southern analysis suggesting that CpG motifs upstream of the promoter region are not methylated in the absence of cytosine methylation at this site.

It is demonstrated from these studies that both $H3K9$ di-and tri-methylation correlate with the DNA methylation and gene expression while $H3K4$ tri-methylation is associated the highest gene expression stage of the *Brachyury* gene. A previous report from Santos-Rosa is in agreement with these data showing that active genes are associated with $H3K4$ tri-methylation in yeast.

This data indicated the same results as for the *Oct4* gene, in that heterochromatin is forming as differentiation occurs again coinciding with the silence of *Brachyury* gene expression.

Effect of TSA on Stem Cell Differentiation

Leukemia inhibitory factor (LIF) was removed from all the cell lines. LIF inhibits cell differentiation, and its removal allows the cell lines to go through cell differentiation. The cell lines were treated with Trichostatin A (TSA) - a histone deacetylase inhibitor for 6 days. One group of cell lines was treated with 10nM of TSA. The western analysis showed the lack of initial deacetylation on Day-1 which, was observed in the control for the embryonic stem cell differentiation. The lack of histone deacetylase activity allowed the acetylation of H3K9 and histone H4. Embryonic stem cells were also analyzed morphologically to observe the formation of embryoid body formation as one of the measures of cell differentiation. The 10nM TSA treated cells failed to form the embryoid body by Day-6 as observed in the control cell line. This implies that the ES cells treated with TSA lacked the deacetylation on Day-1 and failed to differentiate after the removal of LIF. Second group,'-TSA Day4' was treated with TSA for 3days. As soon as the TSA treatment was stopped, on day 4 the deacetylation was observed and the acetylation recovered on Day-5. The morphological examination showed the formation of embryoid body formation by Day-6. In addition, "Interestingly" the embryoid body formation was faster than the control cell line. This suggests that the '-TSA Day4' lines were responding to the removal of LIF but, were unable to acquire any differentiation phenotype. They were able to acquire the differentiation phenotype after the cessation of TSA treatment and at rapid rate. The morphological examination of the third group,' 5 nM TSA' showed the intermediate effect between the control and 10nM-TSA group. The lower dose of TSA allowed the formation of some embryoid body formation. This experiment shows that TSA inhibits histone deacetylase and the activity of histone deacetylase is required for the embryonic stem cell differentiation. Without the initial deacetylation on Day-1, the ES cells cannot go through the differentiation.

Alkaline Phosphatase Activity

In normal stem cells, the activity of alkaline phosphatase activity is lowered upon differentiation. Trichostatin A causes the cells to maintain the activity of alkaline phosphatase. Significant increase in alkaline phosphatase extinction was observed when Trichostatin A was withdrawn after three days. Alkaline phosphatase activity correlates with the morphology changes. Initial deacetylation of histone is required for embryonic stem cell differentiation.

HDAC1, But Not HDAC2 Controls Differentiation

Dovery et al. (2010) used HDAC knockout mice to demonstrate whether HDAC1 or HDAC2 was important for the embryonic stem cell differentiation. Examination of global histone acetylation in the absence of HDAC 1 showed an increase in acetylation. Global histone acetylation levels were unchanged by the loss of HDAC2. In order to analyze the process of HDAC knockout mouse in detail, the knockout mice embryonic stem cells were used to generate embryoid bodies. It showed that just before or during gastrulation, embryonic stem cells lacking HDAC1 acquired visible developmental defects. The continued culture of HDAC1 knockout embryonic stem cells showed that the embryoid bodies formed became irregular and reduced in size rather than uniformly spherical as in normal mice. Embryonic stem cell proliferation was unaffected by the loss of either HDAC1 or HDAC2 but the differentiation of embryonic stem cells were affected with that lack of HDAC 1. This shows that HDAC1 is required for cell fate determination during differentiation.

The Future

Any disturbance of a stable epigenetic regulation of gene expression mediated by DNA methylation is associated with a number of human disorders, including cancer as well as congenital diseases such as pseudohypoparathyroidism type IA, Beckwith-Wiedemann, Prader-Willi and Angelman syndromes, which are each caused by altered methylation-based imprinting at specific loci.

Perturbations of both global and gene-specific patterns of cytosine methylation are commonly observed in cancer while histone deacetylation is an important feature of nuclear reprogramming in oocytes during meiosis.

Recent studies have revealed that there is an array of different pathways that cooperates with one another in order to bestow proper epigenetic regulation by DNA methylation. Future studies will be needed to further clarify the certain mechanism pathways such as DNA binding proteins, DNA repair and noncoding RNAs serve in order to regulate DNA methylation to suppress differentiation and sustain self-renewal in somatic stem cells in the epidermis and other tissues. Addressing these questions will help extend insight into these recent findings for a central role in epigenetic regulators of DNA methylation in controlling embryonic stem cell differentiation.

Stem Cell Laws

Stem cell laws are the law rules, and policy governance concerning the sources, research, and uses in treatment of stem cells in humans. These laws have been the source of much controversy and vary significantly by country. In the European Union, stem cell research using the human embryo is permitted in Sweden, Finland, Belgium, Greece, Britain, Denmark and the Netherlands; however, it is illegal in Germany, Austria, Ireland, Italy, and Portugal. The issue has similarly divided the United States, with several states enforcing a complete ban and others giving financial support. Elsewhere, Japan, India, Iran, Israel, South Korea, China, and Australia are supportive. However, New Zealand, most of Africa (except South Africa), and most of South America (except Brazil) are restrictive.

Science Background

The information presented here covers the legal implications of embryonic stem cells (ES), rather than induced pluripotent stem cells (iPSCs). The laws surrounding the two differ because while both have similar capacities in differentiation, their modes of derivation are not. While embryonic stem cells are taken from embryoblasts, induced pluripotent stem cells are undifferentiated from somatic adult cells.

Stem cell are cells found in most, if not all, multi-cellular organisms. A common example of a stem cell is the Hematopoietic stem cell (HSC) which are multipotent stem cells that give rise to cells of the blood lineage. In contrast to multipotent stem cells, embryonic stem cells are pluripotent and are thought to be able to give rise to all cells of the body. Embryonic stem cells were isolated in mice in 1981, and in humans in 1998.

Stem cell treatments are a type of cell therapy that introduce new cells into adult bodies for possible treatment of cancer, Somatic cell nuclear transfer, diabetes, and other medical conditions. Cloning also might be done with stem cells. Stem cells have been used to repair tissue damaged by disease.

Because Embryonic Stem (ES) cells are cultured from the embryoblast 4–5 days after fertilization, harvesting them is most often done from donated embryos from *in vitro* fertilization (IVF) clinics. In January 2007, researchers at Wake Forest University reported that "stem cells drawn from amniotic fluid donated by pregnant women hold much of the same promise as embryonic stem cells."

In 2000, the NIH, under the administration of President Bill Clinton, issued guidelines that allow federal funding of embryonic stem-cell research.

Europe

The European Union has yet to issue consistent regulations with respect to stem cell research in member states. Whereas Germany, Austria, Italy, Finland, Ireland, Portugal and the Netherlands prohibit or severely restrict the use of embryonic stem cells, Greece, Sweden and the United Kingdom have created the legal basis to support this research. Belgium bans reproductive cloning but allows therapeutic cloning of embryos. France prohibits reproductive cloning and embryo creation for research purposes, but enacted laws (with a sunset provision expiring in 2009) to allow scientists to conduct stem cell research on imported a large amount of embryos from in vitro fertilization treatments. Germany has restrictive policies for stem cell research, but a 2008 law authorizes "the use of imported stem cell lines produced before May 1, 2007." Italy has a 2004 law that forbids all sperm or egg donations and the freezing of embryos, but allows, in effect, using existing stem cell lines that have been imported. Sweden forbids reproductive cloning, but allows therapeutic cloning and authorized a stem cell bank.

In 2001, the British Parliament amended the Human Fertilisation and Embryology Act 1990 (since amended by the Human Fertilisation and Embryology Act 2008) to permit the destruction of embryos for hESC harvests but only if the research satisfies one of the following requirements:

1. Increases knowledge about the development of embryos,

2. Increases knowledge about serious disease, or

3. Enables any such knowledge to be applied in developing treatments for serious disease.

The United Kingdom is one of the leaders in stem cell research, in the opinion of Lord Sainsbury, Science and Innovation Minister for the UK. A new £10 million stem cell research centre has been announced at the University of Cambridge.

Africa

The primary legislation in South Africa that deals with embryo research is the Human Tissue Act, which is set to be replaced by Chapter 8 of the National Health Act. The

NHA Chapter 8 has been enacted by parliament, but not yet signed into force by the president. The process of finalising these regulations is still underway. The NHA Chapter 8 allows the Minister of Health to give permission for research on embryos not older than 14 days. The legislation on embryo research is complemented by the South African Medical Research Council's Ethics Guidelines. These Guidelines advise against the creation of embryos for the sole purpose of research. In the case of *Christian Lawyers Association of South Africa & others v Minister of Health & others* the court ruled that the Bill of Rights is not applicable to the unborn. It has therefore been argued based on constitutional grounds (the right to human dignity, and the right to freedom of scientific research) that the above limitations on embryo research are overly inhibitive of the autonomy of scientists, and hence unconstitutional.

Asia

China prohibits human reproductive cloning but allows the creation of human embryos for research and therapeutic purposes. India banned in 2004 reproductive cloning, permitted therapeutic cloning. In 2004, Japan's Council for Science and Technology Policy voted to allow scientists to conduct stem cell research for therapeutic purposes, though formal guidelines have yet to be released. The South Korean government promotes therapeutic cloning, but forbids cloning. The Philippines prohibits human embryonic and aborted human fetal stem cells and their derivatives for human treatment and research. In 1999, Israel passed legislation banning reproductive, but not therapeutic, cloning. Saudi Arabia religious officials issued a decree that sanctions the use of embryos for therapeutic and research purposes. According to the Royan Institute for Reproductive Biomedicine, Iran has some of the most liberal laws on stem cell research and cloning.

Americas

Brazil

Brazil has passed legislation to permit stem cell research using excess in vitro fertilized embryos that have been frozen for at least three years.

United States

Federal law places restrictions on funding and use of hES cells through amendments to the budget bill. In 2001, George W. Bush implemented a policy limiting the number of stem cell lines that could be used for research. There were some state laws concerning stem cells that were passed in the mid-2000s. New Jersey's 2004 S1909/A2840 specifically permitted human cloning for the purpose of developing and harvesting human stem cells, and Missouri's 2006 Amendment Two legalized certain forms of embryonic stem cell research in the state. On the other hand, Arkansas, Indiana, Louisiana, Michigan, North Dakota and South Dakota passed laws to prohibit the creation or destruction of human embryos for medical research.

During Bush's second term, in July 2006, he used his first Presidential veto on the Stem Cell Research Enhancement Act. The Stem Cell Research Enhancement Act was the name of two similar bills, and both were vetoed by President George W. Bush and were not enacted into law. New Jersey congressman Chris Smith wrote a Stem Cell Therapeutic and Research Act of 2005, which made some narrow exceptions, and was signed into law by President George W. Bush.

In November 2004, California voters approved Proposition 71, creating a US$3 billion state taxpayer-funded institute for stem cell research, the California Institute for Regenerative Medicine. It hopes to provide $300 million a year.

President Barack Obama removed the restriction of federal funding passed by Bush in 2001, which only allowed funding on the 21 cell lines already created. However, the Dickey Amendment to the budget, The Omnibus Appropriations Act of 2009, still bans federal funding of creating new cell lines. In other words, the federal government will now fund research which uses the hundreds of more lines created by public and private funds.

Canada

In March 2002, the Canadian Institutes of Health Research announced the first ever guidelines for human pluripotent stem cell research in Canada. The federal granting agencies, CIHR, Natural Sciences and Engineering Research Council, and Social Sciences and Humanities Research Council of Canada teamed up and agreed that no research with human IPSCs would be funded without review and approval from the Stem Cell Oversight Committee (SCOC).

In March 2004, Canadian parliament enacted the Assisted Human Reproduction Act (AHRA), modeled on the United Kingdom's Human Fertilization and Embryology Act of 1990. Highlights of the act include prohibitions against the creation of embryos for research purposes and the criminalization of commercial transactions in human reproductive tissues.

In 2006, Canada enacted a law permitting research on discarded embryos from in vitro fertilization procedures. However, it prohibits the *creation* of human embryos for research.

On June 30, 2010, The Updated Guidelines for Human Pluripotent Stem Cell Research outline that:

1. The embryos used must originally have been created for reproductive purposes

2. The persons for whom the embryos were created must provide free and informed consent for the unrestricted research use of any embryos created, which are no longer required for reproductive purposes

3. The ova, sperm, nor embryo must not have been obtained through commercial transactions

Canada's National Embryonic Stem Cell Registry:

- contains all human embryonic stem cell lines generated using CIHR funds or funds from any of the research councils

- is a prerequisite for obtaining CIHR funding for human embryonic stem cell research

- will minimize the need to generate large numbers of cell lines, and decrease the need for donation of large numbers of embryos

Oceania

Australia is partially supportive (exempting reproductive cloning yet allowing research on embryonic stem cells that are derived from the process of IVF). New Zealand, however, restricts stem cell research.

References

- Hans R. Schöler (2007). "The Potential of Stem Cells: An Inventory". In Nikolaus Knoepffler; Dagmar Schipanski; Stefan Lorenz Sorgner. Human biotechnology as Social Challenge. Ashgate Publishing, Ltd. p. 28. ISBN 978-0-7546-5755-2.

- Binder, Marc D.; Hirokawa, Nobutaka; Uwe Windhorst, eds. (2009). Encyclopedia of neuroscience. Berlin: Springer. ISBN 978-3540237358.

- Lodish, Harvey (2000). Molecular Cell Biology (4th ed.). New York: W. H. Freeman. Section 14.2. ISBN 0-7167-3136-3.

- Kumar, Rani (2008). Textbook of Human Embryology. I.K. International Publishing House. p. 22. ISBN 9788190675710.

- D. Binder, Marc; Hirokawa, Nobutaka; Windhorst, Uwe (2009). Encyclopedia of Neuroscience. Springer. ISBN 3540237356.

- Stocum DL (2004). "Amphibian regeneration and stem cells". Curr. Top. Microbiol. Immunol. Current Topics in Microbiology and Immunology. 280: 1–70. doi:10.1007/978-3-642-18846-6_1. ISBN 978-3-540-02238-1.

- Knisely, Karen; Gilbert, Scott F. (2009). Developmental Biology (8th ed.). Sunderland, Mass: Sinauer Associates. p. 147. ISBN 0-87893-371-9.

- Engler, AJ; Sen, S; Sweeney, HL; Discher, DE (August 2006). "Matrix Elasticity Directs Stem Cell Lineage Specification". Cell. 126: 677–689. doi:10.1016/j.cell.2006.06.044. PMID 16923388. Retrieved 2016-04-22.

- Serrano, Manuel (2013-09-11). "Study published in Nature is another step towards regenerative medicine" (PDF). cnio.es. Retrieved 2013-12-11.

- Choi, Charles. "Cell-Off: Induced Pluripotent Stem Cells Fall Short of Potential Found in Embryonic Version". Scientific American. Retrieved 25 April 2013.

Permissions

Index

www.ingramcontent.com/pod-product-compliance
Lightning Source LLC
Chambersburg PA
CBHW062001190326
41458CB00009B/2932